Nuclear Winter

Nuclear Winter

THE HUMAN AND ENVIRONMENTAL
CONSEQUENCES OF NUCLEAR WAR

Mark A. Harwell

With Contributions by

Joseph Berry Doria Gordon
Herbert D. Grover Christine C. Harwell
Steven Pacenka David Pimentel

With a Foreword by Russell W. Peterson

Springer-Verlag
New York Berlin Heidelberg Tokyo

Mark A. Harwell
Cornell University
Ecosystems Research Center
Corson Hall
Ithaca, New York 14853/USA

Library of Congress Cataloging in Publication Data
Harwell, Mark A.
  Nuclear winter.
  Bibliography: p.
  Includes index.
  1. Nuclear warfare—Environmental aspects.   I. Title.
U363.H37   1984   355'.0217   84-22126

With 27 Figures

Media conversion by Ampersand Publisher Services, Inc., Rutland, Vermont.
Printed and bound by R. R. Donnelley & Sons, Harrisonburg, Virginia.
Printed in the United States of America.

9 8 7 6 5 4 3 2 1

ISBN 0-387-96093-7 New York Berlin Heidelberg Tokyo
ISBN 3-540-96093-7 Berlin Heidelberg New York Tokyo

*To Kimberly and Matthew,
my commitment to the future*

Foreword

In 1982, three conservationists in the United States discussed a growing concern they shared about the long-term biological consequences of nuclear war; they wondered what such a war would do to the air, the water, the soils, the natural systems upon which all life depends. I was one of those three; the others were executives of two philanthropic foundations, Robert L. Allen of the Henry P. Kendall Foundation and the late Robert W. Scrivner of the Rockefeller Family Fund. Together we began trying to find out what the scientific community was doing about the problem and what steps could be taken to alert the environmental movement to the need to address the subject. We knew that a large-scale nuclear war might kill from 300 million to a billion people outright and that another billion could suffer serious injuries requiring immediate medical attention, care that would be largely unavailable. But what kind of world would survivors face? Would the long-term consequences prove to be even more serious to humanity and survival of all species than the immediate effects?

We found that comparatively little scientific research had been done about the environmental consequences of a nuclear war of the magnitude that today's huge arsenal could unleash. The Hiroshima and Nagasaki blasts were produced by isolated, single fission bombs that each had explosive power equivalent to 13 to 20 thousand tons (kilotons) of TNT, whereas an average thermonuclear weapon today might have a yield of 500 kilotons, and some are as powerful as 20 million tons (megatons). The current worldwide nuclear arsenal consists of about 50,000 nuclear weapons with an aggregate yield of nearly 15,000 megatons; these weapons are capable of reaching targets throughout the world within minutes.

We soon learned that several scientists were indeed looking into aspects of the long-term consequences of nuclear war. A special issue on the human and ecological consequences of nuclear war published in

June 1982 by *Ambio*, the international environmental journal of the Royal Swedish Academy of Sciences, included an especially relevant article indicating an element not previously considered. Dr. Paul J. Crutzen of the Max Planck Institute for Chemistry in Mainz, Federal Republic of Germany, and Dr. John W. Birks of the University of Colorado presented calculations on the extent to which large quantities of smoke and soot from forest fires and burning cities resulting from a nuclear war could produce a thick smoke layer in the atmosphere that would drastically reduce the amount of sunlight reaching the Earth's surface. Their calculations suggested that the darkness could persist for many weeks, rendering agricultural activity in the Northern Hemisphere virtually impossible if the war occurred during the growing season.

We also learned from Dr. Carl Sagan of Cornell University about a study he and four other American scientists were working on that involved this same effect of dust and soot particles and their impact on the world's climate. They had become interested in the subject in 1971 while studying data sent to Earth by Mariner 9, the first spacecraft to orbit Mars. There were severe dust storms on Mars at the time, and it led them to start thinking about the prolonged lower temperatures these dust clouds were producing on the surface of that planet. But it was not until a decade later that Sagan and Drs. Richard P. Turco, Owen B. Toon, Thomas P. Ackerman, and James B. Pollack began to relate the Mars phenomenon to potential consequences of nuclear war. Their comprehensive two-year investigation of the atmospheric effects of nuclear war became known as the TTAPS study (derived from the initials of the authors' surnames).

Using computer models, the TTAPS team analyzed several dozen nuclear war scenarios and investigated their effects on the global atmosphere and climate. The baseline case was a 5,000 megaton war in which about a third of the existing world arsenals would be exploded over a combination of military, urban, and industrial targets, mostly in the Northern Hemisphere. The air bursts over cities could ignite flammable materials and cause vast fires producing sooty smoke; ground bursts could produce enormous clouds of fine dust; and nuclear warhead-induced forest and grassland fires could cause huge plumes of soot.

Smoke plumes would merge to block the sun's light, alter the heat balance, and disturb climate on a global scale. This could cause inland temperatures to drop well below freezing, even in summer. The amount of sunlight reaching the ground, possibly as little as one percent, could be insufficient for plant photosynthesis. The temperature difference could drive the fine particles in the Northern Hemisphere across the equator, bringing climatic changes to the Southern Hemisphere, even if no nuclear bombs exploded there. After a few months or longer, most of the particles would settle out of the atmosphere, but the returning sunlight would include intense ultraviolet light transmitted to the ground through the ozone layer, partially destroyed as the largest nuclear fireballs extended into the stratosphere. Settling cloud particles would bring additional radioactivity to most of the Northern Hemisphere, spreading to areas far from the targets. This combination of prolonged cold, darkness, radioactive fallout, and ultraviolet light could present a significant threat to the survivors of a nuclear war, that is, a *nuclear winter*.

A committee which Allen, Scrivner, and I formed decided that if the TTAPS atmospheric conclusions proved to be substantiated by other expert scientists, and if sufficient additional material could be obtained on the potential biological consequences of the nuclear winter theory, we would convene a conference to present the findings to the scientific and environmental communities and to public officials and other leaders of the public and private sectors. A Conference Steering Committee was established, expanding our group to include physical, biological, and environmental scientists as well as environmentalists and other interested persons.

In late April 1983, fifty scientists met at the American Academy of Arts and Sciences in Cambridge, Massachusetts, to consider and evaluate the preliminary draft of the TTAPS study. The physical scientists generally agreed with the TTAPS conclusions as to the potential for substantial reductions in the amount of solar light reaching the Earth's surface and for the severe climatological changes. The group also discussed stresses such as radiation exposure from fallout, exposure to ultraviolet radiation, and the problem of toxic gases released by combustion of synthetic materials. The findings of the physical scientists were then presented to a group of forty biological scientists to consider the potential impacts of the post-nuclear conditions on the Earth's life support systems. They were told that in the baseline case, average land temperatures might drop to $-25°$ Celsius ($-13°$ Fahrenheit) within a week or two, except in coastal areas; that inland lakes and reservoirs could freeze; that daytime light levels could be reduced by 95% or more; and that these conditions could exist for months. They also considered the effects from long-term exposure to ionizing radiation and ultraviolet light and other effects of the post-nuclear war world. The biologists considered the harm that these potential conditions would do to agriculture, marine, freshwater, and terrestrial systems. A consensus was reached among the biologists that global systems would be in considerable danger from a nuclear winter, and widespread species extinctions and unprecedented human deaths could ensue.

A series of events followed the scientific meetings at Cambridge: The TTAPS group revised their study in response to the peer review and prepared an article (published in *Science*, December 1983) that detailed the atmospheric and climatic consequences of nuclear war. A companion article on the long-term biological consequences of nuclear winter was co-authored by 20 of the scientists attending the biological meeting, written primarily by Drs. Paul Ehrlich, Mark Harwell, Carl Sagan, and George Woodwell. While these articles were being written, the Conference Steering Committee completed its preparations for public presentation of these studies. In addition, Dr. Harwell was asked by the Committee to prepare a detailed technical support document addressing the issues discussed in the *Science* biological article. The preparation of that draft was supported in part by funds from the Conference Committee to the Center for Environmental Research at Cornell University.

In Washington, D.C., on October 31, 1983, the momentous findings about nuclear winter first were publicly presented at the World After Nuclear War, the Conference on the Long-Term Worldwide Biological Consequences of Nuclear War, where it was revealed that even in a war involving only a small part of the world's nuclear arsenals, the horrendous deaths and devastation from immediate blast, radiation, and other effects would be only the first phase of destruction, and could be followed by a climatic catastrophe spreading sub-zero temperatures and a shroud of darkness over most of both hemispheres for a period of several months to a year. The potential results of such a nuclear winter were given national and international news attention for their revelation that the results of nuclear war would likely be far more disastrous than previously believed. Dr. Sagan suggested that the most striking and unexpected consequence of the TTAPS study was that "even a comparatively small nuclear war can have devastating climatic consequences, provided cities are targeted."

More than 500 participants, including scientists and ambassadors or other officials from more than 20 countries, as well as public officials, educators, students, environmentalists, and religious, civic, business, philanthropic, foreign policy, and arms control leaders took part in the Conference. It was sponsored by 31 national and international scientific, environmental, and population organizations or institutes, including the International Union for the Conservation of Nature and Natural Resources (IUCN), United Nations Environment Programme (UNEP), United Nations University, Canadian Nature Federation, Ecological

Society of America, Environment Liaison Centre, American Institute of Biological Sciences, National Audubon Society, Sierra Club, Friends of the Earth, National Wildlife Federation, Natural Resources Defense Council, Wilderness Society, Federation of American Scientists, and World Resources Institute.

The Steering Committee limited the Conference discussion to the physical, atmospheric, and biological consequences of nuclear war, because we felt that the inclusion of other considerations such as disarmament or economic, social, and political implications would distract attention from the central scientific message. The non-political and international nature of the conference was demonstrated by the inclusion in the program of Georgiy S. Golitsyn, Senior Scientist of the Institute of Atmospheric Physics, USSR Academy of Sciences; Dr. Vladimir V. Aleksandrov, head of the Climate Research Laboratory at the Computing Center of the USSR Academy of Science; and Dr. Paul Crutzen from the Federal Republic of Germany.

It was a truly remarkable assemblage and, for many participants, marked a turning point toward realization of the urgent need to convey to the highest policy makers in their countries an understanding of the nuclear winter threat. Immediately following the Conference, participants remained for an even more remarkable companion event, a historic first of its kind. In a live satellite hookup between Washington, D.C., and Moscow, named "The Moscow Link," Dr. Thomas Malone, former Foreign Secretary of the U.S. National Academy of Sciences, moderated a 90 minute discussion on nuclear winter issues among high-level Soviet and American scientists.

Considering the tensions existing between the political leaders of the nations whose nuclear weapons are targeted at each other, it was tremendously encouraging to watch the scientific experts of both sides calmly discuss the consequences of nuclear war, agree on most basic points, and offer additional information from their own research. This collegial atmosphere was summarized by Georgiy Skryabin, Principal Scientific Secretary of the USSR Academy of Sciences:

> On the one hand, there is the feeling of great concern about the possible tragedy that we are facing, that is hovering over all of us—over children, women, old people, and all life on Earth. On the other hand, there is also something that is very pleasing about this Conference, and that is the fact that the great scientists who are sitting here—our American colleagues and Russian scientists—have reached a consensus. They are unified in their views that there should be no nuclear war, that this would mean disaster and death for mankind. I personally am pleased and comforted by this because in our

time the authority of scientists is very great, and we should all try to bring our influence to bear in order to bring about an end to the arms race so that there never will be a nuclear war.

The findings did not prove to be as controversial as some people had expected, although there remains some uncertainty about specific quantitative predictions of long-term climate change, and substantiation is particularly being sought for the projection that the climatic effects of a Northern Hemisphere war would spread to the Southern Hemisphere and could affect global climate for many years after a nuclear war. The TTAPS study and the biological assessments discussed at the Conference and published in *Science* met with general approval in the scientific community. The scientists involved in these studies are the first to recognize that a great deal of scientific research still needs to be undertaken to understand better the potential climatic effects and biological consequences of nuclear war. No one wants to verify the accuracy or inaccuracy of the TTAPS study and the biological paper through the only absolute proof—an actual nuclear war.

Thus, we need books like this one and more. From the draft Dr. Harwell initially prepared as technical support to the *Science* article, he has added considerable material on the immediate effects of nuclear war. With the assistance of several other scientists, he has expanded the book to include analyses of the longer-term human and societal consequences that would develop as the support functions normally performed by human and natural ecosystems ceased to operate in the aftermath of a nuclear war. The present book constitutes an integration of all of these facets of the consequences of nuclear war.

Additional scientific research is already underway, addressing some of the uncertainties about climatic changes and their biological effects, and the quantity and type of nuclear explosions that might trigger the nuclear winter phenomenon. The U.S. National Academy of Sciences has conducted a two-year study, funded by the U.S. Nuclear Defense Agency, on the atmospheric effects of nuclear war, which should be released before the end of 1984. Another major, multi-year study is being conducted by the Scientific Committee on Problems of the Environment (SCOPE), a part of the International Council of Scientific Unions. The SCOPE project is assessing both the climatic and atmospheric changes and the effects of nuclear war on environmental and human systems. This involves the collaborative effort of scientists from the United States, Canada, the United Kingdom, Australia, Japan, the Soviet Union, and several other countries in Eastern and Western Europe and the Third World, with Dr. Harwell coordinating the analyses of effects on agricultural and ecological systems. Nuclear winter was also the subject of symposia at the 1984 annual meetings of the American Association for the Advancement of Science and the Ecological Society of America. Recently, other studies of climatic effects have begun under auspices of the U.S. National Oceanic and Atmospheric Administration (NOAA); and the Lawrence Livermore National Laboratory in Berkeley, California, which designs nuclear weapons, is also investigating phases of the nuclear winter issue.

In an effort to spread information about nuclear winter to the public, bring it to the attention of policy makers, and encourage additional scientific research, the Steering Committee which organized the 1983 Conference established The Center on the Consequences of Nuclear War, in Washington, D.C. The Center has prepared films, videotapes, slide shows, and other educational materials. The effort to inform the public and policy makers of the world on nuclear winter should be a high priority subject for action by individuals in every country. All people, even those living in the most remote parts of the world, are as threatened by nuclear winter as are citizens of the United States and the Soviet Union.

Dr. Lewis Thomas, Chancellor of the Memorial Sloan-Kettering Cancer Center, has said that the nuclear winter discovery "may turn out, in a world lucky enough to continue its history, to have been the most important research findings in the long history of science." In concluding remarks at the Conference, Dr. Walter Orr Roberts emphasized that the scientific work must continue because the issues are not yet fully resolved:

> But we already know enough of the risks to recognize that it is imperative, in the name of humanity, to accelerate the search for world security in the policy domain. As citizens of our own nation states, and as residents of "Spaceship Earth," we must indeed invent and enact policies that convenant a stable future for the planet, and for its pragmatists, poets, saints, soldiers, and indeed for all living, sentient beings.

None of us can responsibly sit back today and let somebody else worry about this problem.

Russell W. Peterson
President, National Audubon Society
Chairman, Center on the Consequences of Nuclear War

Preface

This book is an examination of the consequences of nuclear war. Many studies have focused on the immediate effects of nuclear detonations, often with careful attention given to the two nuclear weapons that were used in war in the destruction of Hiroshima and Nagasaki. We attempt to go beyond such studies by looking at effects not only directly from the nuclear detonations themselves, but also effects occurring over longer time periods and via indirect mechanisms. The intent is to provide a more comprehensive view of what the world would look like to the immediate survivors of a nuclear war.

The approach taken here utilizes scenario and consequence analyses, in which a hypothetical nuclear war is identified in scenario analyses, and the effects of this scenario are examined systematically. The particular emphasis here is on large-scale nuclear wars, i.e., those nuclear wars in which sufficient detonations occur for effects to be widespread. As the number of detonations increases, effects from individual explosions begin to accumulate, so that qualitatively new types of consequences would result. Such new effects include widespread assaults on the environment and the systems that support human civilization.

In order to characterize the scenario, we reviewed many previous studies and selected the exchange of about 5,000 megatons of nuclear warheads on the major industrialized nations as the base case. This scenario reflects attempts by each side to inflict major damage on military, industrial, and civilian targets. For the United States, the scenario includes the detonation of multiple warheads on the Standard Metropolitan Statistical Areas (SMSA's), that is, about 300 cities with populations down to about 100,000 inhabitants. Similar targeting for other countries is assumed. Smaller scenarios could be envisioned, and the current arsenals are such that the total yield of nuclear detonations could be increased two- or three-fold. Thus, the scenario analyzed here is considered to be representative of the suite of potential scenarios of a concerted nuclear war; we do not offer it as the most likely scenario to occur, but, rather, as representing the mid-range of large-scale nuclear war. The scenario is specified in sufficient detail so that consequence analyses can be performed. By conducting some parametric analyses (i.e., deviating from the base scenario for specific parameters), we can understand the sensitivity of the results to the specific assumptions used in defining the scenario, including the magnitudes and distribution of nuclear weapons on various targets.

The first step in these consequence analyses is the characterization of the state of the world in the immediate aftermath of a large-scale nuclear war. By carefully defining the human, physical, and biological conditions immediately resulting from nuclear war, we have the bases for analyses of subsequent time periods. In quantifying these initial conditions, we rely on the studies of the effects of individual nuclear detonations and calculate the effects from the projected exchange.

Results from the initial consequence analyses are that one-half to three-quarters of the U.S. population could be immediate casualties (i.e., dead or injured). Similar analyses elsewhere suggest worldwide deaths of one billion people, with another billion sustaining major injuries. The detonations would result in the deposition of lethal doses of early fallout over a quarter or more of the continental areas of the U.S. and Europe. Further and key to the longer-term consequences, the nuclear detonations could initiate large-scale fires in urban and natural areas, resulting in the injection of massive quantities of dust and soot into the atmosphere.

Analyses performed by physical scientists studying the long-term atmospheric consequences of nuclear war (Turco et al., 1983a,b, 1984) have projected that the atmospheric particulate loadings would lead to major global climatic alterations. In particular, their estimates are for sunlight to be attenuated by the dense coulds of smoke for weeks and months after the nuclear war, with levels of incident sunlight reaching as low as 1% of normal. Consequently, the atmospheric temperatures are expected to undergo drastic reductions, with minimum values after a few weeks of −20°C or lower for average air temperatures in mid-continental areas of the Northern Hemisphere (compared with a current average annual temperature of +13°C). Clearly the consequences on atmospheric systems, if accurate, would presage stresses on humans and the environment of unprecedented scale. This potentiality is one major focus of the present study.

Given the conditions at the end of the immediate period of the nuclear war, we have performed consequence analyses into the longer term by systematically considering the effects on humans and the environment. One result is clear: Human societal systems would be plunged into disarray, as the support functions normally performed by human systems ceased to operate. This would include not only targeted nations, but also areas well removed from direct detonations, as the economic, agricultural, transportation, and other systems failed. A critical result is that humans would be forced to rely on natural systems for food, shelter, warmth, and other essentials. Yet the natural systems themselves would be subject to major stresses that would decrease their capacity for support. Just at the time when greatest reliance on natural systems would occur, these systems would be least capable of providing sustenance. Thus, the analysis of the *ecological* effects of a nuclear war is essential, not to know what the state of the natural world would be, but precisely to define what the *human* consequences could be.

In analyzing the environmental consequences, we present a number of problems, any one of which could have catastrophic effects. Foremost is the effect of temperature reductions directly on humans and especially on the productivity of agricultural and natural systems. We conclude that essentially all terrestrial productivity, including crop production, could be shut down for the first year after a nuclear war, with the obvious implication of human starvation on a massive scale in both hemispheres of Earth. Light reductions also would be expected to result in substantially reduced productivity. Local and global fallout would reach lethal levels over substantial portions of the Northern Hemisphere, and background radiation everywhere could approach levels causing chronic health effects. Finally, societal impacts could also lead to widespread human consequences, as tremendous competition for resources would ensue for the survivors, who would also face physical and psychological stress and the loss of societal support functions.

This report also briefly discusses other types of environmental effects. One of the amazing aspects of nuclear war is that essentially every environmental problem facing the world today would be a part of the effects of a nuclear war, including among many others ozone depletion, habitat destruction, and air and water pollution.

We discuss recovery processes in order to understand the forces acting towards and against the recovery of human civilization. A critical issue here is that human recovery could not proceed more rapidly than the recovery of natural systems, and that the recovery rates of the latter would be controlled and retarded by exploitation by humans having few self-sustaining systems. This feedback between humans and the environment would have much to say about the quality of human existence over the decades and beyond after a large-scale nuclear war.

A synthesis of the various factors that could harm humans indicates the potential for at least as many human casualties in the intermediate- and long-term periods as from the direct effects of the detonations themselves. These indirect fatalities would occur in non-targeted nations as much as the combatant countries.

Estimates of actual fatalities contain substantial uncertainty. However, the stresses projected to occur on humans and the environment are so extreme that the qualitative results are more certain. We conclude the potential for billions of human deaths from intermediate- and long-term effects.

In fact, conditions could be so extreme that we believe there is a real possibility that no humans would survive in the Northern Hemisphere. This is not to say that we predict this would necessarily occur; but for the first time it appears humans could effect such a result. Further, depending on a few uncertain parameters associated with inter-hemispheric transport in the atmosphere, it is conceivable that this risk could be exported to the Southern Hemisphere as well. In such a situation, the potential for the extinction of *Homo sapiens* becomes a valid question. In the event humans do survive, it is clear that the effects of a large-scale nuclear war would continue into the far future. The conclusion is that a large-scale nuclear war constitutes not just war waged among combatant nations, but actually constitutes war waged on all peoples on Earth, war waged on the global environment itself, and war waged on all members of all foreseeable generations.

Acknowledgments

This work represents the collective efforts of many individuals. Although I wrote most of the material in this book, several colleagues also contributed to the initial text. Specifically: Dr. Joseph Berry, staff scientist at the Carnegie Institution of Washington, Stanford, California, prepared initial material on temperature and light reduction effects on biota; Ms. Doria Gordon, Marine Biological Laboratory (currently at University of California, Davis) provided initial text on blast effects and on the effects of low temperatures on human health, and she assisted in many of the calculations; Dr. Herbert D. Grover, University of New Mexico, wrote on fire development and effects on natural systems; Ms. Christine C. Harwell, an attorney with expertise in environmental law and policy, provided overall coordination of the text and wrote the section on societal effects; Mr. Steven Pacenka, Center for Environmental

Research, Cornell University, prepared the discussion of temperature effects on surface water systems; Dr. David Pimentel, Professor of Entomology, Cornell University, prepared the section on agricultural impacts in Chapter 4 (pages 121–130).

In addition to those contributors, advice and information were given by Dr. Wendell P. Cropper (University of Florida), Dr. Jim Detling (Colorado State University), Dr. Barry Edmonston (Cornell University), Dr. Charles Geisler (Cornell University), Dr. John R. Kelly (Cornell University), Dr. Simon A. Levin (Cornell University), Professor Joseph Rotblat (University of London), Dr. Carl Sagan (Cornell University), and Dr. David A. Weinstein (Cornell University).

All of these individuals reviewed and commented on earlier drafts of this document. Additional reviews and comments were provided by Dr. Thomas Eisner (Cornell University), Ms. Karin Limburg (Cornell University), Dr. Walter Lynn (Cornell University), Dr. Thomas F. Malone (Holcombe Research Institute), Dr. Walter Orr Roberts (University Corporation for Atmospheric Research), and Dr. Thomas C. Hutchinson (University of Toronto).

Assisting me in performing calculations were Doria Gordon, Herbert D. Grover, Christine C. Harwell, and Lee F. Timm. Technical support was provided by Nadine Bloch and Tony Whitman. David Weinstein performed the FORNUT simulations, with assistance from Linda Buttel. C.L. Hanson conducted the SPUR simulations. Word processing and administrative support were provided by Carin Rundle, Jacquie Wright, and Colleen Martin.

This report was initially prepared in part as a technical support document for the Conference on the Long-Term Biological Consequences of Nuclear War, held in Washington, D.C., on 31 October 1983. Thus, the conclusions herein represent the consensus of over forty ecologists and biologists associated with the Conference as reflected in the article prepared by the Biological Committee (Ehrlich et al., 1983). Errors in calculations, factual material, and logic are strictly my responsibility.

Ithaca, New York   *Mark A. Harwell*
31 July 1984

Contents

1

Introduction

During the last few decades, humans have developed and implemented the technology capable of causing self-destruction on a totally unprecedented global scale. The initial development and the dramatic use of nuclear weapons at the end of World War II startled the world into a new awareness: that a single weapon delivered by a single vehicle could instantaneously destroy most of the population and the physical structure of an urban area. The horror of such massive casualties was greatly enhanced by the terror of radiation as the unseen killer, striking down its victims not only quickly but also over subsequent weeks, years, and even across generations.

Ironically, though, the limited nuclear war on Japan does not give us an adequate picture of the consequences of a contemporary nuclear war. This is largely a function of scale. For example, the nuclear bombings of Hiroshima and Nagasaki were not *quantitatively* different in terms of human casualties from other single-event but conventional urban bombings of the war; e.g., the 13 February, 1945 raid on Dresden killed 100,000 people, and 85,000–100,000 were killed in Tokyo in the 9 March, 1945, raid (Mark, 1976; Ishikawa and Swain, 1981), levels comparable to the 70,000–140,000 fatalities each of Nagasaki and Hiroshima (U.S. Strategic Bombing Survey, 1946; Ishikawa and Swain, 1981; Barnaby and Rotblat, 1982). Thus, the scale of the numerical consequences of these nuclear detonations versus the maximum utilization of conventional strategic bombing did not differ. Further, the major mechanisms of inflicting injury and death were in large part not *qualitatively* different: The effects of physical trauma from blasts, heat, and secondary fires were the primary causes of casualties for both the nuclear attacks (Ishikawa and Swain, 1981) and conventional fire-bombings. Radiation was, of course, an important agent for injury and death in the Japanese bombings, accounting for less than 5% of the fatalities (Mark, 1976). However, both detonations over Hiroshima and Nagasaki were air bursts, at heights about 500 m above the ground (Barnaby and Rotblat, 1982), such that the fireball never directly contacted the ground. Consequently, local fallout was very small compared to the potential amount from a surface burst. In short, the major new qualitative difference in the nuclear attacks on Japan was strictly a change in the scale for the delivery systems; it was not the number of casualties that could be inflicted in a few hours' time; it was the ease and economy with which it could be done.

Obviously, the point here is not to denigrate the importance to humanity of that first use of nuclear weapons, nor to belittle the unique, profound suffering imposed on the victims. Rather, it is to make clear that the current predicament for human society results from the qualitative changes in the nuclear arsenals of the world that have occurred *since* Hiroshima and Nagasaki. These changes involve massive alterations of the scales of both the explosive energy of individual nuclear weapons and the numbers of such weapons. Individual yields have increased from the approximately 20 kilotons (kT) of the initial weapons to tens of megatons, i.e., an increase by three orders of magnitude; and instead of a few bombs, there are now tens of thousands of strategic and tactical warheads. A weapon the size of the one detonated over Hiroshima is now relegated to minor tactical roles, for example as a depth charge targeted against a single submarine or as a shell fired from field artillery (Center for Defense Information, 1982).

The *quantitative* leap is clear: Today there could be 1,000,000 Hiroshima's (Barnaby, 1982). The *qualitative* change in the nature and effects of nuclear war, though, is not so easily characterized. Concomitant with the vast numerical increases in yields and warheads is a variety of new factors that could define what the world would be like after a major nuclear war. For example, unlike the situation in Japan, after a major strike on the United States civilian population, there would not be the influx of food, medical support, and other aid from non-affected areas into targeted cities; the organized support functions of American society would largely cease to function as all cities became targets and refugia were eliminated. Another example: The magnitude of potential detonations could lead to major changes in light and temperature regimes for long periods of time, and indirect effects on humans from exposure and from loss of agricultural productivity could be very extensive. Further, the large number of strategically important ground bursts could lead to continental-scale radiation fallout at clinical and lethal levels. These and many other such effects reflect the collective consequences of many nuclear detonations, where entirely new mechanisms emerge that threaten *Homo sapiens* and natural ecosystems on a global scale.

That is precisely the raison d'etre of this book, that the tremendous numerical growth in nuclear technology has produced the potential for global consequences qualitatively different from anything previously experienced, yet those consequences have been seriously ignored or inadequately studied and understood. Analysts have typically taken the estimate for the casualties from one nuclear detonation over one city, usually only those casualties resulting from blast effects, and multiplied this by the number of blasts per city and the number of cities targeted to calculate the consequences of a major nuclear war. The resulting estimates, of tens of millions of casualties, are of course exceedingly large and reflect the potential for unprecedented death and injury. For example, a single warhead detonated over a single U.S. city could inflict more fatalities than all the American deaths in all wars, in battles from Concord through Danang. Yet using *only* such an approach can tremendously underestimate the actual consequences. The other direct effects and especially the indirect effects that emerge as the scale increases to a large number of nuclear detonations can be as severe as or even greater than the effects of blast. Further, by looking only at the latter, there results a dangerously flawed picture of life after such a nuclear war: flawed in the sense of suggesting survivors can just go about the business of their lives with only short-term perturbations on the functioning of society [e.g., the recent reassurance from the U.S. Postal Service that mail will be forwarded to evacuees of targeted cities within weeks (Miller, 1982)]; dangerous in the sense of suggesting that the post-nuclear war world would be readily habitable by those who survive the immediate time period and, therefore, that nuclear war *is* thinkable and does not prima facie constitute organized, societal suicide.

Underestimating the full range of consequences leads to a lack of appreciation of the state of the world after a nuclear war in those areas and countries that are not participants in the war. In a war involving the United States, Soviet Union, and Europe, not only would there be millions of survivors in those countries in the immediate period, even under the most extreme of scenarios, but there would be $2-3 \times 10^9$ (i.e., billion) people in the other countries of the world that would be well removed from the direct effects of the detonations themselves. By looking only at casualties from blast, those non-involved countries could develop a false sense of security and thereby grossly ignore the vital interest they have in such an event not occurring. As we shall see, no country is safe from potential catastrophe; clearly understanding that fact is extremely important to the community of nations on Earth.

Other problems result from looking only at direct effects: It can lead to a perception that limited nuclear war, such as involving a counterforce strike on selected strategic targets, is feasible and perhaps acceptable in terms of civilian impacts. Such a scenario, in the range of 1000–2500 megatons (MT) total yield of detonations over the United States (Katz, 1982) could be specified with civilian casualties in the range of 2–20×10^6 (million) deaths in the first thirty days (OTA, 1979), in this case primarily from the effects of local fallout. Again, we shall see that actually the longer term and indirect effects would extend well beyond the ranges of local fallout plumes, and they would in fact engulf the entire hemisphere in extreme stress conditions.

A similar situation relates to the consequences of a genuine first strike by one side on the other, where the aggressor receives no retaliatory warheads. Analysis of only the direct effects would indicate that the aggressor would experience no casualities at all, except for the relatively slight increase in long-term cancer rates from global fallout. Decision makers could be led to conclude that such a scenario would constitute a tremendous victory by inflicting catastrophic damage to the military and societal systems of the other side but with no substantial cost to their own side. Yet a comprehensive analysis of the consequences of such a scenario shows that an enormous toll of human casualties from indirect effects, at levels totally unacceptable under any criteria, would be taken on each side, including the one with no detonations in its half of the world. As elsewhere, clear understanding of this assessment by decision makers could completely change the nature of strategic policy. Why work toward the technology of a first strike capability if its successful implementation would still constitute self-destruction, involving many tens of millions of casualties in one's own country?

Returning to the issue of the quality of life for survivors after a nuclear war, it is essential to evaluate the longer term and indirect consequences in order to understand the processes and impediments for recovery of the beneficial physical, societal, and biological environments required for human existence. This has major implications for the nature and efficacy of civil defense policies; for predicting the forces acting toward or against reestablishment of social order and organized political, cultural, and economic systems; the rates of recovery and the stability of agroecosystems and the gradual increase in the carrying capacity of the planet for human populations; and at a more local level, the factors that would control the lives and define the personal environment of each survivor and his or her family.

In summary, the consequences of nuclear war beyond just the spatial and temporal extents of blast zones involve a variety of factors that would have overriding importance in establishing the numbers of survivors in targeted and nontargeted countries alike, the extremes of physical stresses placed on their lives, and the quality or adversity of the lives of every individual on Earth for decades and generations after a nuclear holocaust. At the scale of a nuclear war against cities and military bases across whole continents, war is also waged on the global environment itself, allowing no refuge. It is essential that such an understanding be generally held; perhaps the strongest deterrent to large-scale nuclear war is an accurate conceptualization at all levels of the social and political hierarchy of what it would invoke.

There have been other examinations of the longer term and indirect effects of nuclear war, particularly including studies by the National Academy of Sciences (NAS, 1975), the Office of Technology Assessment (OTA, 1979), the Hudson Institute (Ayres, 1965), the United Nations (UN, 1981), the Arms Control Disarmament Agency (ACDA, 1979), Katz (1982), and a series of analysts writing for a special issue of *Ambio* (Ambio Advisory Group, 1982) (see also Grover et al., in prep.).

By reviewing these previous studies, consistent trends are evident: (1) the recognized scope of potential mechanisms for adverse effects has expanded over time as new areas of potential effects were identified; (2) the issues of scale, including scales of the number of nuclear detonations, the spatial extent of impacted areas versus refugia areas, and the temporal sequence of differing factors controlling effects were not adequately treated; and (3) a single issue typically was identified as the most important mechanism for deleterious effects, but that issue differed across the studies (e.g., initially it was the immediate direct effects of the bombs, then local and global radioactive fallout was emphasized, followed by particular concern over the ozone (O_3) depletion and associated ultraviolet (UV-B) enhancement, and most recently the consequences from climatic alterations induced by massive atmospheric inputs of particulates from fires). The historical lesson is that the more the consequences have been examined, the greater in both extent and intensity are the results; there has been a consistent underestimation of the total consequences of nuclear war, even by those scientific studies explicitly addressing the longer term and indirect effects.

The purpose of this book is to provide a reasonably comprehensive look at the consequences of a nuclear war large enough to trigger the more profound long-term and broad-scale effects on humans and the environment. These analyses are prepared as a technical support document for discussions on the biological aspects of nuclear war as presented at the Conference on the Long-Term Consequences of

Nuclear War (31 October–1 November 1983, Washington, D.C.) and as outlined in the report representing the consensus of the biologists and ecologists affiliated with that Conference (Ehrlich et al., 1983). This work is closely associated with a similar study on the atmospheric effects of nuclear war, as discussed in Turco et al. (1983a) and at the Conference. In particular, the atmospheric analyses followed the initiative of Crutzen and Birks (1982) in evaluating the changes in light levels, air temperatures, and radioactive fallout induced by multiple nuclear detonations and concomitant injection into the atmosphere of large quantities of particulates from widespread fires in urban and natural areas.

The calculations of Crutzen and Birks (1982) and Turco et al. (1983a) (essentially confirmed by Covey et al., 1984; Alexandrov and Stenchichov, 1983; and MacCracken, 1983) have convincingly raised the issue of potentially drastic reductions in temperatures for substantial periods of time in response to many-fold reductions in the transmissivity of the atmosphere. The scale of these reductions may reach hemispheric or even global proportions. Thus, the specter is raised of global casualties of humans and other species, sufficient to warrant this current look at these and other environmental stresses. Additionally, the expansive growth in the world's nuclear arsenals over the last several years and the current level of friction among the nuclear powers justify this examination of one potential future for human civilization.

Underlying these analyses is a three-part theme that will be developed in this work:

1. The scale of effects from both direct and long-term insults on humans and the environment is such that human casualties would be extraordinary in extent. Most of the societal systems that support human populations at levels currently well above the carrying capacity of the unmanaged biosphere would be severely curtailed or terminated. This includes agroecosystems at least throughout the Northern Hemisphere, the health care systems at least in all targeted countries, the energy and transportation systems in developed countries, the economic systems of all countries, and the societal systems of most or all countries. In short, modern civilization would at least temporarily cease to function in a recognizable way. No longer could an individual or a nation rely on ordered society to provide the services essential to its survival.

2. Because of this loss of human support systems, there would be an enforced reliance on the support functions of natural ecosystems. As they found themselves without adequate food, shelter, energy, medical care, transportation, communications, and other necessities, survivors would demand support from natural systems in order

simply to survive. This includes those people in overpopulated and underdeveloped countries for whom subsidies from the developed countries in the Northern Hemisphere (i.e., those countries most likely to be directly involved in the nuclear conflict) currently provide the difference between subsistence and widespread famine. Throughout the world, immediate survivors would have to escape their current urban habitats and seek new means of self-support by exploiting the entire landscape.

3. However, acting in direct contravention to the latter, the nuclear war would result in extensive, adverse effects on natural ecosystems, significantly reducing the capacity of the environment to support humans and other species. Just at the time when an impotent human society would be forced to place a greatly increased reliance on natural systems, the very bases of ecosystem productivity and stability would be undermined.

It is the nexus of these themes that establishes the ecological effects of nuclear war as central to the survival of the world's human population and as determining the quality and recovery of civilization. The fate of *Homo sapiens* would be literally left to the vagaries of Nature reeling from unprecedented, anthropogenic insults. Consequently, concern about the ecological effects of nuclear war is not a trivial matter relegated to environmental extremists, narrow-perspective biologists, or those callous to human suffering. It is simply not correct to have the perspective of why worry about trees and birds, when millions of people are dying. Rather, it is precisely an acute interest in the human condition after a nuclear war that requires an understanding of the ecological effects. Throughout this document, the ultimate focus is on human impacts, and all environmental considerations are within that context.

The structure of the text follows the sequence of scenarios, consequence analyses, and recovery processes. Chapter 2 provides the details and justifications for the base scenario analyzed here and the ranges of values used for parametric studies. From the scenario development process come the characterization of various aspects of a hypothetical large-scale (but not "worst-case") nuclear war in whatever detail is necessary. The analyses are described in Chapters 3 and 4, the former characterizing the initial conditions needed for subsequent analyses (specifically describing the state of the human and environmental systems at the end of the immediate period of the war), the latter using that information for calculations and inferences about the intermediate- and long-term consequences on humans and the biosphere. In this presentation is an emphasis on certain major problem areas, each of which potentially could devastate the human population, followed by lesser problem areas, of significant but relatively less catastrophic importance. Chapter 5 addresses the recovery processes for humans and the biosphere, including feedbacks between the two and the forces acting to impede recovery. Finally, Chapter 6 is intended to recall the essential elements of these analyses and to provide a single synthesis of the state of the Earth after a nuclear war, including some awareness of the pervasive potential for synergistic responses.

2

Scenario Development

The use of scenario/consequence analyses in the study of complex problems has become widespread. By identifying causal relationships between system inputs and system responses, consequence analyses attempt to treat stress responses in a systematic manner, including not just the direct relationships, but also linkages that operate via intermediate processes or components. It is because of the intrinsic complexity of input/system response relationships that often a substantial number of the characteristics of the system and the stresses are required to be explicitly identified and quantified. Scenario analysis is performed in order to accomplish that.

Scenarios, then, constitute the set of information that defines the system and its stresses to the degree necessary for adequate consequence analyses. Scenarios are needed in order to provide the essential framework for consequence analyses; otherwise those analyses may be too abstract and removed from reality, may miss key elements that ultimately could control the system responses, or may be poorly integrated into a comprehensive conceptualization of what the system actually would look like after experiencing the perturbation. Each of these deficiences has been experienced in previous analyses of the consequences of nuclear war.

In the development of scenarios for this study, we have deliberately focused on large-scale nuclear war, since the intent of the analyses is to describe consequences of nuclear war on a continental or global scale, not just the effects from a single or few nuclear detonations. The latter have been well treated (see Glasstone and Dolan, 1977, and Ishikawa and Swain, 1981, for particularly thorough assessments of nuclear weapons in general and the Japanese bombings in specific), so that there is a high degree of confidence in projections of the physical effects of a small number of nuclear detonations. Only by dealing with multiple nuclear detonations do we get into the qualitatively new problem areas, such as

atmospheric and climatic alterations, societal collapse, and global
fallout. As previously pointed out, projections of the total consequences,
including these types of effects, have not been adequately studied.

Another factor in considering large-scale nuclear war scenarios is the
issue of an initially limited nuclear war. It is not our intent to address the
political and military considerations of limited versus large-scale nuclear
war. However, it should be clear what types of scenarios would fall into
our category of large-scale nuclear war. As discussed previously, many
analyses of a "limited" nuclear war involve total nuclear yields detonated
over the U.S. in the few thousand megaton range (Katz, 1982; OTA,
1979), including counterforce strikes on all ICBM silos, strategic air
bases, and other military targets. Such a scenario, as we shall see, in fact
constitutes a nuclear war of magnitude sufficient to trigger massive
global-scale effects, and, thus, is considered to be a large-scale nuclear
war for the purposes of this document. Only those scenarios in the tens
of megatons range are below the scale of consideration here. Such small-
scale, truly limited nuclear wars involving the superpowers, however,
may be very difficult to contain, and an inevitable escalation to a large-
scale war that would have widespread consequences may ensue
(Zuckerman, 1982; Sagan, 1983). Thus, both the plausibility of a small-
scale nuclear war between the U.S. and the U.S.S.R. inevitably escalating
to a scale in excess of 1000 megatons (MT), and the qualitatively new and
profound effects on global human and ecological systems that would
result from such a scale of nuclear war, argue for restricting our analyses
to large-scale nuclear war scenarios. Further, we resist the inclination to
analyze increasingly smaller scenarios as potentially providing the bases
for the definition of what level of nuclear war could be construed to be
"acceptable"; we do not wish the emphasis on new, global consequences
emerging from larger-scale scenarios to suggest that below some
threshold value, the global environment is safe and that the use of that
magnitude of nuclear detonations is a viable, sufficiently benign
option.

In selecting the scenarios for subsequent analysis, sufficient specificity
is required in order for most of the consequence calculations to be
made. It is insufficient, for example, merely to define a nuclear war in
terms of the total yield of the nuclear detonations as the basis for
consequence analyses. The criticism has been raised that this approach
limited the usefulness of the NAS study (1975), since many important
consequence analyses were inadequately treated, largely because of
insufficient specification of initial conditions (Grover et al., in prep.).

The degree of specificity varies with the type of calculation being
performed. For instance, knowing the number of warheads, their various

individual yields, and the heights of detonations as air bursts is adequate for projections of the total warhead-generated oxides of nitrogen (NO_x) into the stratosphere. Other calculations require greater precision of scenario characterization; e.g., accurate projections of local fallout plumes from multiple detonations require knowledge of the number of warheads of each yield, the height of bursts, the climatic conditions (precipitation, wind-patterns, etc.), the local topography, the fusion-fission ratio of the weapons, and the spatial patterns of nearby detonations and concomitant overlap of fallout patterns. In the analyses presented here, the goal was to specify the scenario as necessary and sufficient for consequence analyses, but not with greater specificity. This is to minimize the potential for scenario-specific conclusions and to make the projections more understandable and accessible to decision makers.

The import of this book is not as an evaluation of current and projected nuclear arsenals and military strategies. Thus, we have tried to minimize the level of effort needed for identifying particular war scenarios. In particular, we have surveyed the previous studies for scenarios appropriate for our consideration, including review of NAS (1975), OTA (1979), Katz (1982), UN (1981), Bergstrom et al. (1983), Ayres (1965), Lewis (1979), ACDA (1979), Hill and Gardiner (1979), Mark (1976), Schell (1982), Woodwell (1963), and the *Ambio* study (Ambio Advisory Group, 1982). In general, these contained scenarios that were based on outdated information about military arsenals, contained insufficient specificity for adequate consequence analyses, or were far too small a scale of war (one or a few detonations) for the purposes here.

Another constraint was the need for consistency with the atmospheric and climatic analyses being performed in concert with this work (Turco et al., 1983a). In these analyses, a number of scenarios were analyzed, with a 5000 megaton nuclear war providing the base case. This could not be used directly as the scenario here, since for many elements of importance to human and ecological effects, but not to atmospheric effects, further specifications were required.

An important factor in scenario selection was not to base our analyses on a "worst-case" approach. Such a scenario was presented by Schell (1982), which assumed that all of the strategic nuclear warheads on the Earth were detonated over the maximum possible number of targets, including civilian targeting of cities down to 2000 inhabitants. There are a large number of factors that would preclude that from happening, although in theory there are enough warheads available; these factors include warhead failure, delivery system failure, antiballistic missile systems, warhead interference ("fratricide"), insufficient number of delivery systems, incapacity to control that many firing systems, and effects of electromagnetic pulse. Thus, using a worst-case approach is not defensible as being plausible or even credible, and the associated consequence analysis' credibility would likewise suffer, even if it were appropriately done.

The alternative we chose is a *representative scenario* approach, in which a scenario is developed for a large-scale nuclear war, where each side is trying to effect counterforce and countervalue strikes against the other, rather than trying to detonate all possible warheads. The scenario is taken to be representative in the sense of using logical patterns of warhead assignments to various types of civilian and military targets. The intent is not to offer the scenario as the most likely of all scenarios, nor even as a probable scenario. Rather, there is a full suite of possible scenarios that could be envisioned and that are of the scale needed for global impact analyses; we have sought a scenario in the middle range of that suite. The significance of changing to a scenario with aspects well above or below that middle range is then tested in a series of sensitivity analyses, providing an estimate of the robustness of the conclusions and, consequently, the degree of scenario independence.

A final point is that we have ignored the use of tactical nuclear weapons. Concentrating on large-scale nuclear war dictates primary emphasis on the strategic nuclear weapons of the major nuclear powers. We do not address the consequences of tactical weapons being detonated, because their additive impact globally is not likely to be significant. This is not, however, to suggest that a theater nuclear war in Europe solely involving large numbers of tactical nuclear weapons would not initiate consequences of global impact on humans or the environment; in fact, such a war would fall into our category of being large-scale and causing substantial far-field and long-term effects (see, e.g., Arkin et al., 1982), and many of the alterations in the hemispheric temperature and light regimes would ensue from a scenario of the scale of an extensive European theater war (Turco et al., 1983a).

Considering these factors, we selected the scenario developed for the *Ambio* study (Ambio Advisory Group, 1982) as having by far the greatest detail and relation to plausible military strategies and as being similar in scale with the base case of Turco et al. (1983a). We have followed the *Ambio* scenario in most aspects, deviating only when necessary to evaluate the effects of altering some major assumptions, such as the time of year of the nuclear war. A major advantage of using the *Ambio* scenario is the direct comparability of the present study and the *Ambio* analyses. To a very real degree, our analyses supplement rather than supplant the *Ambio* results, with the particular addition here of the consequences of the major atmospheric alterations projected by Turco et al. (1983a).

A synopsis of the scenario that we developed based largely on the scenario of the Ambio Advisory Group (1982) is presented in Table 1. An equal yield from fission and fusion reactions is assumed. The exchange is assumed primarily to involve North America, Europe, and the Soviet Union, with detonations on other countries as appropriate for military or political purposes. Three categories of targets are assigned: civilian population centers; civilian industrial centers; and military targets. The population targets are assumed to include all cities over 100,000 in most of the Northern Hemisphere. For detailed analyses of the direct civilian casualties in the United States, we modified the *Ambio* scenario to include all of the Standard Metropolitan Statistical Areas (SMSA's) (U.S. Bureau of the Census, 1982); this includes about 300 urban areas.

Table 2 details the suite of warheads assigned to each size class of cities. This follows the *Ambio* scenario, except that for cities greater than three million in size, the number of warheads is pro-rated according to the population size (e.g., a city of six million would have twice the number of each size of warhead assigned to the three million city). Also unlike the *Ambio* scenario, the detonations over the urban areas were assumed in our base case to be all air bursts at optimal height (calculated in accordance with Glasstone and Dolan, 1977), maximizing the immediate deaths from blast and thermal radiation; this assumption was changed in some parametric analyses, with analyses performed for all surface bursts (maximizing local fallout); for a 50/50 mixture of surface and optimal height air bursts, the consequences would be approximately half-way in between the calculated values.

Targets in the industrial and military categories closely follow the *Ambio* scenario. The distribution of weapons among these is listed in Table 3. The specific analyses done for the immediate and local fallout consequence analyses for the United States were based on military

Table 1. Synopsis of Base Scenario.[a]

Targets	Number of warheads (worldwide)	Yield of warheads (MT)
Cities	5000	1950
Military	6600	3000
Industrial	3150	700
Total	14750	5650

[a]Scenario developed by author following scenario of Ambio Advisory Group (1982).

Table 2. Warheads Detonated Over Urban Centers.[a]

Population size	Number of warheads	Yield of warheads	Type of burst
10^5–10^6	3	300 kT	Air
	1	100 kT	Air
1–3×10^6	3	1 MT	Air
3×10^6	10	500 kT	Air
	5	1 MT	Air
$>3 \times 10^6$	Pro-rated		Air

[a]Scenario developed by author following scenario of Ambio Advisory Group (1982).

targets identified in the maps prepared by the Ambio Advisory Group (1982) (Figure 1), with the addition of the 300 U.S. SMSA's. Also added into our scenario are nuclear detonations on nuclear power facilities, dams, hazardous material storage areas, and other targets that could have complicating outputs.

The nuclear war is assumed to occur with minimal warning and with minimal duration, so that the population is not assumed to be different from normal distribution patterns. Reflecting census data, the war is to occur, in effect, while people are in their residences; the effect of a war during times of increased density in central cities, as during typical working hours, is addressed in parametric analyses.

The issue of failure of delivery and warhead systems is avoided by considering that the listed warheads actually detonate, constituting in yield less than one-half of the potential strategic arsenal available (SIPRI, 1982b,d); we do not try to assign a numerical value to the numbers of

Table 3. Warheads Detonated Over Industrial and Military Targets.[a]

Target	Number of warheads	Yield of warhead	Type of burst
Military bases	1	300 kT	Surface
Early warning centers	1	100 kT	Surface
Submarine detection bases	1	300 kT	Water
ICBM bases	2	500 kT	Surface
Submarine bases	1	1 MT	Surface
Large industrial complexes	5	200 kT	Air
Oil fields	2	500 kT	Air
Energy facilities	1	10 kT	Air

[a]Scenario developed by author following scenario of Ambio Advisory Group (1982).

Figure 1. Map of the military targets of the United States as defined in Ambio Advisory Group (1982) scenario. The large symbols represent ICBM bases. The scenario in this book also includes all of the SMSA's, and selected industrial, energy, and other targets.

warheads that are launched but for some reason fail to reach the appropriate target and/or fail to detonate.

In contrast to the Ambio Advisory Group (1982) selection of a particular hour on a particular June day, the analyses here do not assume a specific time of day, time of year, or particular weather patterns. We discuss the effects of seasonality wherever appropriate, and variations in local weather conditions are ignored in calculations such as for initial fallout. Other environmental parameters were specified as necessary to effect specific consequence analyses. These and other detailed assumptions will be made apparent in the discussions of the consequences.

That discussion begins in the next chapter. Presented there are the initial conditions of what the world would be like at the end of the immediate time period (within a few days after the nuclear exchanges).

3

Initial Conditions

In this chapter, we provide detailed information needed to describe the state of the world at the end of the immediate post-nuclear war period. This is the first aspect of consequence analyses, as we calculate the effects on human and ecological systems that would occur during or immediately after the war. These numbers are necessary for two reasons: first to detail what the immediate cost of such a war would be and provide a basis for comparison with other analyses, many of which are limited to only these immediate consequences; and second to provide all the information necessary in order to perform the analyses of the subsequent consequences on humans, society, and the environment. In accordance with the latter point, this chapter will also include some analyses extending in time beyond the immediate period but necessary as precursors to other, longer-term analyses. For instance, included here will be a discussion of the state of the atmosphere with respect to particulate, radioisotope, photooxidant, and other loadings, and with respect to the resultant atmospheric and climatic changes predicted in the Turco et al. (1983a) work. Those results which include, for example, temperature depressions occurring maximally several weeks after the war and extending into the few-year time frame, then, are essential initial conditions for the subsequent consequence analyses of the effects on humans and the environment discussed in later chapters. Thus, *initial* conditions here include but are not limited to *immediate* conditions.

Human Health Effects

This section provides the essential information to enumerate human casualties realized in the immediate time period during and following the scenario war. These represent direct effects only, although all direct effects will not be included; e.g., long-term cancer induced by global fallout is a direct effect but not included in these immediate-term consequence analyses. The calculations presented here are estimates of human fatalities and injuries. For these immediate, direct human casualties, three distinct mechanisms will be analyzed: blast, thermal radiation, and initial ionizing radiation. For each of these, there is a fairly well established relationship between the yield of each detonation, the intensity of the effect, and the distance from ground zero. The basic approach for each is to calculate the areal coverage of damage for each warhead yield assigned in the scenario to each size class of city in the United States, overlay that onto the population associated with the city class, and calculate the fraction of the population affected. Calculations for other countries are not included here; Bergstrom et al. (1983) provides estimates for worldwide casualties.

Effects from Blast

The fission or fusion reactions in nuclear weapons release tremendous quantities of energy in a very small volume and over a very short period of time, resulting in extreme local increases in temperature (up to tens of millions of degrees C) and pressure (up to millions of atmospheres) (Glasstone and Dolan, 1977). The nuclear reactions are completed within the span of less than 10^{-6} second (i.e., one millionth of a second); thus, the dramatic developments of the fireball, the mushroom cloud, and the associated release of energy as heat, radiation, and blast all occur well after nuclear fission and/or fusion are over.

Initially, the extremely high temperatures cause a release of radiant energy in the form of thermal X-rays, which are absorbed rapidly by the immediately surrounding atmosphere, resulting in almost instantaneous heating and reradiation at slightly longer wavelengths from the molecules in the surrounding air. In this manner, the fireball develops and expands into a luminous, spherical mass of air and debris from the warhead (Glasstone and Dolan, 1977). This rapid expansion of an air mass creates a high-pressure shock wave as outside, cooler air is displaced; that shock wave travels initially at supersonic speeds and radiates out in all directions from the fireball. When the shock front contacts the ground, it is reflected back into another shock wave front. But this reflected shock wave travels through air that has already been compressed and heated by the initial incident wave, so that its velocity is somewhat greater than the incident wave. Consequently, at a particular distance on the ground from the blast (approximately equal to the height of the air burst), the reflected wave catches up with the incident wave (known as the "Mach" effect), reinforcing its intensity and expanding the area covered by a certain level of destruction. Further, the merging of the two waves results in a new wave (the "Mach stem"), which is virtually perpendicular to the ground surface, thereby creating associated winds that are parallel to the surface, substantially increasing the effective destructive capability of the blast wave (Glasstone and Dolan, 1977).

As the Mach stem passes a point on the surface, the local air pressure will instantaneously jump from normal ambient to a greatly elevated level, then gradually decay back to normal, and even to below pre-blast ambient levels. The initial change in pressure is termed the *peak overpressure* above ambient atmospheric pressure, typically measured in pounds per square inch (psi). The level of the peak overpressure is closely associated with the degree of damage experienced from a blast, based on extensive studies from nuclear weapons tests and from the Japanese bombings. Affiliated with the overpressures of the blast wave are dynamic pressures, resulting in very strong winds following passage of the blast wave itself (see Glasstone and Dolan, 1977, for a thorough analysis of blast effects).

The unprotected human body is rather resistant to the effects of overpressure: Some humans can survive a blast with peak overpressure of 30 psi, and the LD_{50} value for the human body is about 12 psi (Middleton, 1982). (LD_{50} refers to the level that would be lethal to 50% of the individuals in an exposed population.) However, buildings and other structures are not nearly as resistant, and widespread failure of buildings occurs at levels of just a few psi (Glasstone and Dolan, 1977). Thus, the human casualties from blast are almost totally caused indirectly by blast destruction of structures (e.g., from collapsing buildings and violently flying debris).

Empirical evidence indicates that the fatality rate for humans in urban areas at the 5 psi boundary is about 75%, with, of course, higher rates closer to the blast and lower rates further away. The area around the blast that includes peak overpressures $\geq$ 5 psi is considered to be the *lethal area* (Lewis, 1979; Barnaby and Rotblat, 1982), where the lethal area is defined to include about as many survivors within it as there are fatalities outside the area. That is, in effect the outside fatalities can be considered to be moved into the area in exchange for the inside survivors being removed from the lethal area. This greatly simplifies calculating the number of fatalities from a blast by merely counting the population within the 5 psi isopleth.

It should be understood that the 5 psi value reflects a single nuclear detonation; multiple detonations could be expected to act synergistically. For example, the structural damage out to, say, the 2 psi distance from one blast could make the area much more subject to damage and resultant human casualties from a subsequent blast wave, probably approaching from a different direction, so that its 2 psi area could become lethal (Katz, 1982).

In calculating the effects of blast at a particular distance, the warhead yield and height of the burst need to be considered. Ideally, a specified overpressure will occur at a distance that is proportional to the cube root of the energy yield, and empirical evidence supports this relationship for yields up to 1 MT (Glasstone and Dolan, 1977). Therefore,

$$\frac{D}{D_{ref}} = \left[\frac{W}{W_{ref}} \right]^{1/3} \tag{1}$$

where D = slant range (line of sight) distance from the detonation
        W = yield of warhead (kT)
        ref = reference warhead values.

By choosing $W_{ref} = 1$ kT,

$$D = D_{(1kT)} \times W^{1/3} \tag{2}$$

This applies only to detonations occurring at the same scaled height as the reference detonation:

$$h_s = h_{ref} \, W^{-1/3} \tag{3}$$

where h_s = scaled height
        h_{ref} = reference height.

Further, above 5000 ft (1500 m) altitude of detonation, the atmosphere can no longer be treated as being homogeneous, and corrections for pressure and temperature deviation from the reference warhead must be made.

Glasstone and Dolan (1977) provide sets of curves for a free air burst of 1 kT (i.e., a burst at sea-level atmospheric pressure that does not contact the surface or other boundary). These curves show the peak over-pressures experienced at various distances from reference free air bursts that occur at various heights above the ground. From these relationships, an optimum height is apparent for creating the maximum overpressure at a particular distance; this therefore corresponds to the height at which the area covered by a particular overpressure is the greatest. Also, a surface burst (in which part of the fireball comes in direct contact with the ground) is seen to have a considerably reduced area covered by a specified overpressure in comparison with an air burst, largely because much of the energy is expended in excavating the ground, volatizing solids, etc., rather than contributed to a blast wave.

Using these curves and the appropriate scaling factors for warhead yield, we calculated the optimum heights for each size warhead used in the base case scenario, and then calculated the radial horizontal distance to which at least 5 psi overpressure would be experienced. These results are shown in Table 4, along with the distances to the 2 psi isopleth, taken to be representative of the area of substantial, nonlethal physical injury to humans (Katz, 1982).

From the suite of warheads assumed to be detonated over each size class of city in the United States (Table 2), calculations were made of the lethal areas for each city in order to make estimates of numbers of fatalities. The cities were treated as two separate groups: the Standard Metropolitan Statistical Areas (SMSA's) as categorized in the table; and the ten largest cities, for which more careful attention was given. The latter were treated separately because of their large size and the wide variance in density gradients in the surrounding metropolitan regions.

For the SMSA grouping, the total lethal area for the base case was assumed to be the sum of the lethal areas for each warhead for each city size class. This assumes no overlap among the circles of effect, a reasonable assumption, since each city only has three or four warheads, but an assumption that somewhat overestimates the blast-induced fatalities. This calculated total area was converted into a single effective circular area, and its radius was calculated. This allows placement of a single circle over the center of the city, maximizing the fatalities by

Table 4. Blast Parameters.[a]

Warhead yield (W)	Optimal height (km)[b]	5 psi Radius (km)		2 psi Radius (km)	
		Air	Surface	Air	Surface
100 kT	1.56	3.28	2.11	5.64	3.54
200 kT	1.95	4.16	2.67	6.77	4.35
300 kT	2.24	4.72	3.04	8.06	4.99
500 kT	2.65	5.60	3.65	9.67	5.96
1 MT	3.35	6.72	4.56	12.41	7.57

[a]Calculations performed following Glasstone and Dolan (1977).
[b]$h = 1100W^{0.45}$.

selective placement over the greatest population density areas. However, it was also assumed that there were no synergisms among the warheads (i.e., that the full 5 psi value is needed to define the lethal area); this potentially underestimates the casualties. As a variation on the base scenario, these calculations were also performed for the 5 psi isopleth for the various size classes of cities but from surface bursts. This latter scenario would represent the minimal effects from blast but the maximal effects from local fallout, an alternate possible strategy for targeting. All areas for the SMSA cities are shown in Table 5.

The areas impacted by $\geq$ 5 psi peak overpressure from blasts on the largest ten U.S. cities were calculated similarly, using the suite of warheads assumed in the Ambio Advisory Group (1982) scenario for a city of 3 million people, but pro-rated to the actual size of each city as determined from 1980 census data (U.S. Bureau of the Census, 1983). The population of each city was calculated by including the central city and all associated suburbs out to a region where the population density approached a uniform background level for that region. The SMSA's that fell within the major city designation were then subtracted from the SMSA classification, so that no individuals were double counted. A specific number of warheads was assigned to each city, as shown in Table 6. Areas out to the 5 psi isopleth for each warhead type were calculated as before, and a total area for each city was determined, again assuming no interactions among detonations and, for the upper estimate, assuming no overlap. These values constitute the base estimate, since the latter two assumptions oppose each other in their effect on casualty estimations. However, overlap of blast areas was considered to be more of a problem for the largest cities, since many warheads were assigned to each.

Table 5. Calculated Blast Areas.

Population (city size class) (in thousands)	5 psi Peak overpressure			
	Areas (km^2)		Radii (km)	
	Air burst[a]	Surface burst	Air burst[a]	Surface burst
100	243.8	101.0	8.81	5.67
100–250	243.8	101.0	8.81	5.67
250–500	243.8	101.0	8.81	5.67
500–1000	243.8	101.0	8.81	5.67
1000	425.7	195.9	11.64	7.89

[a]Optimal height.

Table 6. Ten Largest U.S. Cities Data.

City	Population[a]	Nuclear detonations (numbers)			
		1 MT	500 kT	300 kT	100 kT
Boston	3,667,900	6	12	1	
Chicago	7,806,600	13	26		
Dallas	2,772,700	4	7	1	1
Detroit	4,429,900	7	15	1	
Houston	2,868,400	5	9		
Los Angeles	11,318,800	19	37	1	
New York	11,511,400	19	38	1	1
Philadelphia	5,199,000	8	18	1	
San Franciso	4,850,400	8	16		2
Washington, D.C.	2,826,300	5	8	1	1
Total	57,251,400				

[a]Population figures derived from 1980 census data for the central cities and associated surrounding communities (US. Bureau of the Census, 1983).

Therefore, an estimate of overlap was made to provide a lower value for casualties.

The maximum overlap possible would occur if each warhead had precisely the same epicenter. However, that does not seem a likely military strategy for destroying urban areas, since it would constitute a considerable waste of delivered warheads. Rather, it was reasoned that an enemy would attempt to distribute the warheads to achieve maximum effectiveness; yet, trying too hard to avoid overlap would lead to reduced casualties as warheads would be dispersed to areas of lower population density. Therefore, we assume that the maximum area without overlap would be sought, as calculated previously, but that within that area the warheads would be randomly located. Calculations were made of the resultant reduction in areas covered by 5 psi overpressures as follows (S.A. Levin, personal communication):

Let A_i = area of 5 psi circle from ith warhead, and $\Sigma A_i = A_T$ = maximal total area covered by blast.

$$\Sigma a_i = 1 \quad \text{where} \quad a_i = \frac{A_i}{A_T}. \qquad (4)$$

Then the probability that a point within A_T is covered by a single, randomly placed circle is

$$p_i = a_i$$

and the probability of that point not being covered is

$$q_i = (1 - a_i).$$

The estimate of the area within A_T that is not covered by at least one circle, assuming independence of placement of circles is:

$$\phi = \prod_{i=1}^{n} (1 - a_i) \qquad (5)$$

where ϕ = non-covered area
n = number of circles.

In the situation where all circles are the same size,

$$\phi = (1 - a)^n. \qquad (6)$$

But $a = 1/n$ by definition, so

$$\phi = (1 - 1/n)^n. \qquad (7)$$

Since

$$e^z = \lim_{m \to \infty} \left(1 + \frac{z}{m}\right)^m \qquad (8)$$

then Eq. 7 reduces to

$$\phi \cong e^{-1} \text{ for large } n. \qquad (9)$$

It can be seen that this holds for mixed size circles as well. For example, assume two sizes of circles; then

$$\phi = (1 - a_1)^{n_i}(1 - a_2)^{n_2} \qquad (10)$$

and

$$a_1 n_1 + a_2 n_2 = 1. \qquad (11)$$

If $b = a_i n_i$ then

$$b_1 + b_2 = 1 \text{ and} \qquad (12)$$

$$\phi = \left(1 - \frac{b_1}{n_1}\right)^{n_1} \left(1 - \frac{b_2}{n_2}\right)^{n_2} \qquad (13)$$

$$\cong e^{-b_1} e^{-b_2} = e^{-1}. \qquad (14)$$

This can be shown for additional size classes of circles, and it can be shown that the value for ϕ converges on e^{-1} quickly as all n_i increase; further, these conditions do not have to be met very precisely for the approximation to be accurate.

Given this, the lower estimate for total area covered, but by randomly overlapping circles, is given by

$$A_T(1 - 1/e) \qquad (15)$$

or about a 36% reduction in coverage. This value was used to adjust the bomb coverage on the largest cities. As with the SMSA calculations, another parametric analysis was done by considering coverage by surface blasts rather than the optimum height air bursts of the base scenario. Table 7 presents these data for the ten largest U.S. cities.

To translate the calculated impact areas into human fatalities for the SMSA's, the urban density model of Edmonston (1975) was utilized. In an extensive study of American cities, he showed that the cities of a size class have a consistent pattern in density gradients radiating outward from the central city. The model which best fits this relationship is a single exponential decay function of distance, given as

$$y = ae^{-bx} \qquad (16)$$

Table 7. Ten Largest U.S. Cities —Lethal Areas from Blast.[a]

City	5 psi Areas (km^2)		
	Airburst		Surface burst
	No overlap	Random overlap[b]	
Boston	2088	1230	1171
Chicago	4406	2785	2472
Dallas	1361	860	764
Detroit	2541	1606	1426
Houston	1596	1009	895
Los Angeles	6395	4042	3588
New York	6542	4135	3670
Philadelphia	2978	1883	1671
San Francisco	2779	1757	1559
Washington, D.C.	1601	1012	898

[a]In this table and related tables, more digits are reported than are significant digits. This is so that calculations could be independently confirmed; it is not to imply a particular level of precision.

[b]Area with overlap calculated from Eq. 15.

where y = population per unit area as a function of distance x
        a = central density (individuals per unit area)
        b = coefficient of decay.

Edmonston (1975) also derived the integrated form of this equation, so that the cumulative population out to a certain distance from the city center is given as

$$N_{(x)} = (a\ \theta/b^2)(1 - (1 + bx)e^{-bx}) \qquad (17)$$

where $N_{(x)}$ = population within distance x of center
        θ = circular segment of city (in radians).

The θ correction is for cities that do not constitute a complete circle. For instance, a coastal city like Seattle may be limited to a semicircle ($\theta = \pi$ radians), whereas a city like Atlanta is essentially circular ($\theta = 2\pi$) in its distribution.

Values for a and b were calculated by Edmonston and Guterbock (1984) from demographic data based on over 200 SMSA's using 25 years of census information, up to 1975. From those data we estimated 1980 values by linearly proportional extrapolation of the trends from 1970 to 1975. Estimates of average θ values for each size class of cities were calculated from unpublished data of Edmonston. The parameters for a, b, and θ were used in a computer program solving Eq. 17 for each city size class, with output given in 0.1 mile increments (also listed in metric units).

A limitation of this approach leading to an underestimation of far-field effects, however, is the assumption in Edmonston's model that the population of a city approaches an asymptote as x → ∞, i.e., the density approaches zero at large distances. Actually, the model was used only to predict distances up to the point of reaching the average non-urban population density of the U.S., taken to be 24.7 individuals/km² (U.S. Bureau of the Census, 1983). Solving Eq. 16, we get

$$x_{max} = \left(-\frac{1}{b}\right)\left(\ln\frac{24.7}{a}\right). \qquad (18)$$

Thus, in using the computer output of the exponential model, blast radii less than x_{max} for a city size class were used directly to indicate the total fatalities. If the blast radii exceeded x_{max}, then the calculation was

$$F(r) = N(x_{max}) + 24.7\ (A(r) - A(x_{max})) \qquad (19)$$

where F(r) = fatalities for blast effect of radius r
        A(r) = area to radius r
        $A(x_{max})$ = area out to background density.

The parameters used in the SMSA calculations are listed in Table 8. The estimated fatalities for the SMSA cities from blast effects are given in Table 9A.

In order to estimate the casualties for the largest ten cities, detailed maps of each were inspected carefully in comparison with U.S. census data for 1980 (U.S. Bureau of the Census, 1983). Each city was divided into units of relatively homogeneous population density, so that when areas affected by blast and other effects are overlaid on the map, the population included could be directly calculated by multiplying the area of each unit covered by its average density value, and summations made across units. The specific methodology used is described as follows.

The regions of relatively constant density were determined from color coded maps in the Urban Atlases (U.S. Bureau of the Census, 1974) of the top ten U.S. cities and some nearby cities. For nearby cities not covered in the atlases, the maps of urban areas in the 1980 Number of Inhabitants booklet (U.S. Bureau of the Census, 1982) were used as a guide. These regions are superimposed on the 1980 maps of county subdivisions and places (U.S. Bureau of the Census, 1980), using the reasonable assumption of no change in relative densities between the 1970 census and the 1980 census. These regions are large enough and the range of density in each is broad enough to absorb any error resulting from small changes in the boundaries of small constant density areas over that time period. Furthermore, in Texas and California, boundaries of places change as the population changes, so the regions drawn in these states were altered slightly to take advantage of this fact. To calculate the total population of each region, these maps were used to show what places and county subdivisions are to be included in each region, and then the population figures were found in the appropriate table of the Number of Inhabitants booklets of corresponding states (U.S. Bureau of the Census, 1982). If a region divides a county

Table 8. SMSA Density Parameters.[a]

Population (city size class) (in thousands)	a (density) (persons/mi^2)	b (gradient) (mi^{-1})	θ (Radians)
100	15,772	0.991	6.03
100–250	9,112	0.502	5.59
250–500	8,513	0.332	5.71
500–1000	8,512	0.222	5.33
1000	13,406	0.165	4.84

[a]Data from Edmonston and Guterbock (1983) and Edmonston (unpublished data) for U.S. cities.

Table 9. U.S. Casualties from Blast Effects.

A. *Fatalities for SMSA Cities* (lethal area = 5 psi overpressure)

Population (city size class) (in thousands)	Number of cities[a]			Fatalities per city	Total fatalities
	Total U.S.	Included in 10 cities	SMSA's (difference)		
Air Bursts					
100	250	106	144	90,490	13,030,560
100–250	114	27	87	154,059	13,403,133
250–500	33	3	30	240,283	7,208,490
500–1,000	16	5	11	317,547	3,493,017
1,000	6	6	0	793,710	0
					37,135,200
Surface Bursts					
100			144	83,355	12,003,120
100–250			87	105,963	9,218,781
250–500			30	142,708	4,281,240
500–1,000			11	168,431	1,852,741
1,000			0	462,990	0
					27,355,882

B. *Fatalities for Ten Largest Cities*

City	Fatalities		
	Air burst		Surface burst[c]
	No overlap	Overlap[b]	
Boston	2,204,990	1,393,775	1,102,495
Chicago	6,235,775	3,941,635	3,117,890
Dallas	1,137,135	718,780	568,570
Detroit	3,395,900	2,146,550	1,697,950
Houston	1,260,890	797,010	630,445
Los Angeles	9,632,700	6,088,830	4,816,350
New York	9,623,900	6,083,265	4,811,950
Philadelphia	4,154,300	2,625,935	2,077,150
San Francisco	2,350,250	1,485,595	1,175,125
Washington, D.C.	1,697,849	1,073,210	848,925

[a]Numbers from U.S. Bureau of the Census (1983) for 1980 data, less those SMSA's incorporated in ten largest city calculations.

[b]Areas reduced by e^{-1}; population figures assumed to be reduced in proportion to areal reduction, providing a slight underestimation.

[c]Areas reduced by approximately 0.561 from no overlap airburst figures; population assumed to be reduced by 0.5.

Table 9 *(continued)*

C. *Total Fatalities from Blast*

	Urban population at risk[d]	Air burst		Surface burst
		No overlap	Overlap	
SMSA's	54,884,200	37,135,200	37,135,200	27,355,882
Ten Cities	57,251,400	41,693,689	26,354,585	20,846,850
U.S. total	112,135,600	78,828,889	63,489,785	48,202,732
Fraction of targeted population (%)	100.0	70.3	56.6	43.0
Fraction of U.S. population (%)	48.8	34.3	27.6	31.0

D. *Injuries from Blast[e]* (2 psi isopleth)

	Air bursts
SMSA's	10,445,185
Ten Cities	22,533,800
U.S. total	32,978,985

E. *Total Casualities*

	Injuries	Fatalities	Fatalities and injuries
Fraction of targeted population (%)	29.4	70.3	99.7
Fraction of U.S. population (%)	14.3	34.3	48.6

[d]This is the total population within the top ten cities and the rest of the top 300 SMSA's for the United States. The difference between this value and the 230,000,000 people in the U.S. reflects the population that lives outside the top ten cities as characterized in this study and outside the SMSA's beyond the distance at which the average asymptotic population for each size class is realized. The difference value, then represents those people outside the targeted cities of our scenario.

[e]Injuries were calculated by including all of the population within the 2 psi isopleth and subtracting the population within the 5 psi isopleth (i.e., fatalities). For most cities, the 2 psi isopleth extends beyond the distance at which the background densities are reached (25 per km^2 for SMSA's and 100 per km^2 for the top ten urban areas). Thus, the additional area covered by the 2 psi isopleth was multiplied times the background density, and that figure was added to the population left within the cities but not killed.

subdivision, then a percentage of the population of the subdivision, based on the fraction of the area of the subdivision, was used. If the unit contained a place, then the population of the place was used, and a fraction of the remaining population of the surrounding subdivision, based on the division of the remaining area, was treated separately. This method reduces error because places often contain most of the population of a subdivision. Moreover, places contain less area, so assuming a homogeneous density of population is more accurate. County density figures were used for adjacent counties that did not contain a city, since the range of density inside the county is fairly small in those instances.

This procedure was done for all counties encompassing the metropolitan area. It is felt that this method gives the best estimate of the urban populations associated with the large cities that would actually be vulnerable to a nuclear attack on that city, without the substantial error resultant from using population data based just on city limits and other arbitrary political divisions. Also, we believe the units of assumed homogeneous population density were on a scale small enough so that a reasonably accurate casualty figure could be estimated from the fractional areas covered.

To estimate the fatalities from blast over the ten largest cities, the density maps created as described above were overlain with the area associated with the suite of weapons assigned to each city, as listed in Table 7. This was done by covering the most dense areas first, then proceeding outward from the central city to less dense areas. In this manner, the largest number of inhabitants that the blast area could cover were included, providing an upper estimate of casualties. However, it was felt that the assumption of independence of effects from each detonation could significantly underestimate the casualties. Further, a countervalue targeting strategy would likely be differentially oriented to hit the more dense areas. Therefore, this approach is considered to give the best estimate of fatalities.

The fatality estimates for the largest cities are given in Table 9B, and the total calculated fatalities from blast are shown in Table 9C. More detailed targeting information than available to us would be needed to refine these estimates, by using circles for each specifically targeted location, but somewhat modified for variance associated with the "circular error probable" (CEP) values for the specific warhead systems. Nevertheless, we believe the methodology used here gives an accurate estimate of the expected casualties and provides a closer inspection of the site-specific and city size-specific characteristics that could affect the estimates than in previous published studies (e.g., Ambio Advisory Group, 1982; Bergstrom et al., 1983).

Unlike the situation for fatalities, where the 5 psi isopleth is taken to define the lethal area, blast-induced injuries are not as consistently treated in the literature. Clearly, injuries are more related to flying glass and other debris and to collapsing structures than to the effects of blast on the human body alone, as discussed previously. For calculations here, we assume that 2 psi peak overpressure is sufficient to cause injuries for all people at risk, but no injuries occur at areas beyond that level (following Middleton, 1982, and Glasstone and Dolan, 1977). Obviously, there would be people within the 2 psi area who would escape blast injuries, just as there would be people at, say, 1 psi who would be injured. The assumption is that those two sets of people are about equal in size, and that the 2 psi area (minus the included 5 psi areas of fatalities) constitutes a "injury area" comparable to the lethal area concept.

Calculations for the population subject to 2 psi peak overpressures were performed as described for the 5 psi fatality estimates. However, in many cases the 2 psi areas exceeded the areas associated with the targeted cities themselves, so that some inclusion of exurban population is done here. Table 9D presents the results of the calculations for blast-induced, non-fatal injuries.

Direct Effects of Thermal Radiation

According to the second law of thermodynamics, eventually all of the energy released from a nuclear detonation will be converted to heat, including the energy associated with the blast wave and scatter of weapons debris, the energy in radioisotopes, and the energy released immediately as electromagnetic radiation. The latter includes about three-fourths of the total energy of the detonation and, as we discussed, initially occurs as radiation in the thermal X-ray end of the spectrum when the fireball temperatures approach $10^7°C$. This radiant energy is either scattered or absorbed by contact with matter; absorption of the thermal X-rays occurs rapidly [about 90% is absorbed within 5 cm for X-rays transmitted from the fireball (Glasstone and Dolan, 1977)], heating the molecules in the air to a point where reradiation occurs at slightly longer wavelengths and thereby heating still other molecules. By progressing through a series of these adsorption/reradiation steps, the radiation reaches the infrared (IR) in large part, which is not nearly so absorbed by the molecules in the air but which readily transfers heat to other materials it contacts. This process is the source for the effects of thermal radiation during the first few seconds after a nuclear detonation.

Thermal energy per unit area decreases with distance from the point of emission as a function of radiant energy being spread over a larger spherical surface and with attenuation as it passes through the atmosphere. Thermal energy is emitted uniformly in all directions from its source; therefore, the amount of energy/unit area is inversely proportional to $4\pi r^2$, where r is the distance from the source. Atmospheric attenuation results from absorption and scattering, particularly from atmospheric particulates. Atmospheric transmittance of thermal radiant energy varies according to the path length and opacity of the medium, which is generally expressed in terms of visibility, the distance at which a large, dark body has just enough contrast with the surrounding area to be visible in daylight. On a clear day, visibility is roughly 20 km, whereas light haze reduces visibility to 10 km, and heavy haze reduces it to 4 km. Many calculations of radiant energy exposure at distance from a nuclear weapon are performed assuming visibilities of 20 km, thereby providing upper estimates of effects from thermal radiation (Glasstone and Dolan, 1977). One point about scattering effects on visibility: The energy in the thermal radiation is still there, and still at the same frequency. Increased scattering acts to decrease the direct, line-of-sight radiation received from a point source, but the total background radiation received from all other directions is enhanced by scattering. Thus, the degree of atmospheric visibility does not affect the total received thermal radiation as much as would first appear to be the case.

Surface bursts result in considerably lower levels of thermal radiation compared to air bursts because of shielding by terrain; the absorption of light by the low dust layer resulting from the blast; the considerable dissipation of the energy potentially available for thermal X-ray emission by excavating and volatilizing the ground; the greater density of air near the ground; and the greater absorption and scattering by the much higher levels of carbon dioxide (CO_2) and water near the surface. By contrast, high altitude air bursts do not release much thermal radiation, because the air density is so low that a very large volume is needed to absorb the initial thermal X-rays; a volume so large that incandescence does not ensue as air temperatures are not elevated sufficiently.

The effects of thermal radiation are felt only when it is absorbed. A material transparent to the infrared will not be affected, as is the case for highly reflective materials. The problem arises for absorptive materials, since the incident light is of extremely high intensity, though occurring over but a brief period of time. The energy that is absorbed cannot be transmitted through the absorbing material rapidly enough for dissipation to occur as conductivity values for most materials are too low. Thus, the outer portions of the material attain greatly elevated temperatures; it is this phenomenon that results in burns, scorching, and perhaps ignition. Thicker organic material, including human skin, chars. For many materials, the incomplete combustion of the surface molecules results in a considerable emission of smoke, which acts to absorb subsequent incident IR and dissipate its energy as the kinetic energy of the particles suspended in the air, thereby preventing further damage and ignition of the solid material. Also, the outer skin of the material often literally explodes off the solid, likewise dissipating energy and shielding the solid.

Because of the mechanism by which thermal radiation causes damage, the cumulative energy absorbed does not uniquely characterize the extent of damage: A time component is important also. A higher total absorbed energy is required to cause the same damage if the duration of exposure is increased. In the case of nuclear weapons, the lifetime of the fireball varies with weapon yield, according to the relation (Glasstone and Dolan, 1977):

$$t_{max} = 4.17 \times 10^{-2} \ W^{0.44} \qquad (20)$$

where W = weapon yield (kT)

t_{max} = time to maximum thermal energy production in the second thermal pulse (sec).

Thus, for a 10 kT weapon, the fireball's maximal thermal radiation occurs after about 0.1 sec, whereas for a 1 MT detonation, it is about 0.9 sec, and for 20 MT, 3.25 sec. Associated with this, an increased total adsorbed thermal energy is needed to cause a particular level of damage with increased weapon yield.

To calculate the total radiant exposure of some material at a particular distance from the detonation, assuming no atmospheric attenuation, the energy can be considered to be distributed over the surface of a sphere. The total energy released divided by the area of that sphere having a particular radius (i.e., at a specified distance from the detonation) is the cumulative energy density at that distance:

$$Q = \frac{E}{4\pi r_s^2} \qquad (21)$$

where Q = radiant exposure (cal/cm^2)

r_s = slant range from detonation (km)

E = total energy of weapon (kT).

But light attenuation decreases that by an exponential decay term:

$$Q = \frac{E}{4\pi r_s^2} \, (\exp(-kr_s)) \qquad (22)$$

where k = atmospheric absorption coefficient (function of wavelength).

When scattering is also present, the formula must again be modified. According to Glasstone and Dolan (1977), an empirical formulation is:

$$Q = \frac{E\tau}{4\pi r_s^2} \qquad (23)$$

where τ = transmittance (unitless; the fraction of total radiation that is transmitted after absorption and scattering as a function of distance).

Glasstone and Dolan (1977) also reduce this to the equation:

$$Q = \frac{85.6 \, fW\tau}{r_s^2} \qquad (24)$$

where f = fraction of yield as radiant energy (unitless)
W = yield (kilotons)
Q = radiant exposure (cal/cm^2)
r_s = slant range (kilofeet).

In considering the direct effects of these exposures, three main problems are identified: flash burns; eye damage; temporary flash blindness. The first is the most significant in terms of human fatalities and significant injuries. Second-degree flash burns that cover 30% of the body and third-degree burns over 20% of the body will usually result in death in the absence of intensive medical care (UN, 1980). That is to say, under actual nuclear war conditions where such care would not be available, most serious second-degree burns would be fatal or would contribute to death in concert with other factors.

The actual absorption of thermal radiation by the body, however, is highly variable, since shielding is readily accomplished by even thin material between the victim and the burst. The Japanese experience indicated the dramatic effects of skin and clothing exposure versus being in the shadow of some other object (Ishikawa and Swain, 1981). On the other hand, as the t_{max} values in Table 10 indicate, the time between the initial flash and the maximal thermal pulse is quite brief, so active shielding could not occur. The thermal effects on an individual would depend on the capricious timing and location of the detonation, the

Table 10. Thermal Radiation Characteristics.

A. *Air Bursts*

Warhead yield (kT)	h (optimal[a] height) (km)	Q[b] (cal/cm^2)	f[c]	τ[d]	r_s[e] (slant range) (km)	r[f] (ground range) (km)	t_{max}[g] (sec)
		100% Fatality Values					
100	1.56	8.5	0.35	0.7	4.79	4.53	0.32
200	1.95	8.8	0.35	0.7	6.66	6.37	0.43
300	2.24	9.0	0.35	0.7	8.06	7.74	0.52
500	2.65	9.5	0.35	0.7	10.13	9.78	0.64
1000	3.35	9.8	0.35	0.6	13.06	12.62	0.87
		50% Fatality, 50% Injury Values					
100	1.56	5.5	0.35	0.7	5.95	5.75	0.32
200	1.95	5.7	0.35	0.7	8.27	8.04	0.43
300	2.24	5.8	0.35	0.6	9.30	9.02	0.52
500	2.65	5.9	0.35	0.6	11.90	11.60	0.64
1000	3.35	6.0	0.35	0.6	16.69	16.35	0.87

B. *Surface Bursts*

Warhead yield (kT)		Q (cal/cm^2)	f	τ	r_s (slant range) (km)	r (ground range) (km)	t_{max} (sec)
		100% Fatalities					
100	—	8.5	0.18	0.8	4.56	4.56	0.32
200	—	8.8	0.18	0.7	4.77	4.77	0.43
300	—	9.0	0.18	0.7	5.78	5.78	0.52
500	—	9.5	0.18	0.6	6.73	6.73	0.64
1000	—	9.8	0.18	0.5	8.55	8.55	0.87
		50% Fatalities, 50% Injuries					
100	—	5.5	0.18	0.7	4.27	4.27	0.32
200	—	5.7	0.18	0.6	5.49	5.49	0.43
300	—	5.8	0.18	0.6	6.67	6.67	0.52
500	—	5.9	0.18	0.5	7.79	7.79	0.64
1000	—	6.0	0.18	0.5	10.92	10.92	0.87

[a]$h = 1100W^{0.45}$

[b]From Figure 2 after Glasstone and Dolan (1977).

[c]From Glasstone and Dolan (1977) Tables 7.88, 7.101.

[d]From Glasstone and Dolan (1977) Figure 7.98.

[e]From Eq. 24.

[f]$r = (r^2_s - h^2)^{1/2}$.

[g]From Eq. 20.

position of the victim, and the line-of-sight interruption by objects between the person and the burst. Ironically, whereas there would usually be adequate time to protect oneself from blast before the wave arrived, but no real capability to do so, there is not enough time for active protection from thermal radiation, yet a shield as flimsy as a bedsheet would suffice.

Glasstone and Dolan (1977) show relationships between the radiant exposure for an unprotected population and the amount of second- and third-degree burns to be expected (see Figure 2). The values for particular individuals vary substantially with skin pigmentation; dark-skinned individuals require much less exposure to cause the same level of burn damage as light-skinned persons, since much more of the incident energy would be absorbed. About 50% of the exposed population would receive second-degree burns (the rest first-degree burns) at Q values of 5–6 cal/cm² for 100 kT–1 MT weapons, and about 50% would receive third-degree burns (the rest second-degree burns) at values of 8–10 cal/cm². For our purposes, we assume the latter range of values constitutes an LD_{100} for exposed individuals without medical treatment (comparable to Barnaby and Rotblat, 1982). For the population exposed to thermal radiation between the two ranges, we assume

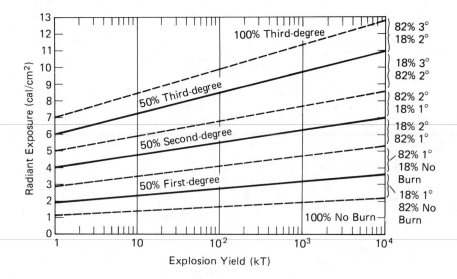

Figure 2. Probabilities of skin burns for average unshielded population. (From Glasstone and Dolan, 1977)

50% mortality, the remainder as serious injuries. Table 10 lists the specified Q levels and associated t_{max} and distance values (from Eq. 20) for warheads of various sizes.

These distances are translated into affected population numbers exactly as the peak overpressure values were used to calculate deaths and injuries from blast. However, the dose/response values discussed above are for persons unsheltered and thereby exposed to the thermal radiation. Those individuals inside and away from windows or otherwise within a shadow from the burst would not suffer any thermal radiant energy effects. It is assumed here that the fraction of the population at risk (i.e., within the critical thermal areas) that actually is exposed is 1% to 33%. The time of day would affect the fraction exposed considerably; e.g., nuclear detonations at 2 a.m. would find very few people exposed, but detonations at rush hour could find a large fraction exposed. Using this information, the estimated numbers of casualties have been calculated, as reported in Table 11.

The second effect of the thermal pulse is damage to the eye. In the Japanese bombings, virtually no permanent eye damage occurred from thermal radiation (Glasstone and Dolan, 1977), largely because the cornea is transparent to most IR and because the burst was probably not in the direct field of vision of most surviving people. Potentially, ultraviolet (UV) damage to the eyes could be severe for those directly viewing a detonation; most of the UV would have been absorbed and reradiated at longer wavelengths along with the rest of the electromagnetic radiation, but only a relatively low level of UV is required for permanent eye damage to occur. Such retinal burns can be localized and not impair vision totally. This phenomenon is discussed further in Glasstone and Dolan (1977), but not considered of major importance here.

Flash blindness is the temporary loss of vision from the extreme intensity of the flash, and it may occur from scattered light as well as direct vision; thus, not looking at the burst may not provide adequate protection. This phenomenon would be substantially enhanced during darkness, when the pupil of the eye would be more open. During the several minutes duration of flash blindness after a burst, useful vision would be lost, incapacitating victims and causing reduced ability to respond to falling structures, fires, etc., during the very immediate period. Glasstone and Dolan's charts (1977) indicate this could be a problem for those within about 30 km on a clear day and 100 km at night from an air burst at 3 km height, relatively independent of weapon yield. Thus, in the scenario of these analyses, a very large population would be subject to flash blindness, often from multiple bursts. In essence, everyone in a targeted city (about 120 million people in the United

Table 11. Casualties from Thermal Radiation.

A. *Population at Risk*

	Population in area 1[a]	Population in area 2[b]	Population at risk for lethal effects	Additional population at risk for injuries
	Air Bursts			
SMSA	9,335,200	12,468,850	10,902,025	1,566,825
Ten cities	22,730,400	28,305,500	25,517,950	2,787,550
Total	32,065,600	40,774,350	36,419,975	4,354,375
	Surface Bursts			
SMSA	14,436,000	16,494,050	15,465,025	1,029,025
Ten cities	17,239,600	19,517,700	18,378,650	1,139,050
Total	31,675,600	36,011,750	33,843,675	2,168,075

B. *Casualties[c]*

	Fatalities	Injuries	Total
Air bursts	1.8×10^5–6.0×10^6	2.2×10^5–7.5×10^6	4.0×10^5–1.4×10^7
Surface bursts	1.6×10^5–5.5×10^6	1.8×10^5–6.2×10^6	3.4×10^5–1.2×10^7

[a]Includes total population within radii defining 100% fatalities for exposed individuals (Table 10) less the population within the 5 psi isopleth (i.e., blast fatalities). As with blast injuries, some areas exceeded the city areas, so 25 or 100 people per km^2 outside the cities were added in for the extra areas.

[b]Calculated as for (1), but for 50% fatality, 50% injury values of Table 10.

[c]The actual casualties from direct thermal radiation would be less than reported in Table 11A since not all people at risk within thermal effect areas would be exposed. Here we assume a range of 1%–33% of the at risk population actually being outside, near windows, in automobiles, etc. Of those exposed within Area 2, we assume half are exposed sufficiently for lethal effects to occur, the other half with significant non-lethal burn injuries.

States) plus millions more within 30–100 km from a military target would be at risk for flash blindness.

Thermal Radiation and Fires

In addition to the direct effects of thermal radiation on humans, indirect effects from thermal radiation- (in combination with blast-) induced fires are very important; in fact, the firestorms created in the Nagasaki and particularly Hiroshima bombings caused a large fraction of the casualties (Bond, 1946; Ishikawa and Swain, 1981).

As in direct effect considerations, the extent and effect of fire following a nuclear detonation are functions of the warhead yield and local atmospheric conditions; larger warheads require greater exposures to cause the same degree of fire initiation. Also important are fuel availability (quality and density), topography, building configurations, weather, vegetation distribution, and surface water systems, both with respect to fire initiation and spread. Low cloud cover or snow coverage on the ground can increase the reflectivity of thermal radiation.

Urban conditions of high building and fuel densities, ruptured fuel lines, disturbed electrical lines, proximity to detonations, and high levels of damage to structures from blast (thereby exposing fuel for combustion) favor extensive fire spread. In contrast to wildland fires, which are discussed in a later section of this book, urban fire ignition and spread can occur under seemingly unfavorable meteorologic conditions. Whereas relative humidity in summer and dewpoint in winter have been found to be good correlatives with ignitability of interior kindling fuels, wind, rain, or snow have not (Chandler et al., 1963). The abundance of interior kindling fuels is considered critical for primary fire ignitions to occur in urban environments (DCPA, 1973; Martin, 1974; Wiersma et al., 1973).

Ignition thresholds of some typical urban kindling fuels are shown in Figure 3 as a function of weapon yield, thermal irradiance, and emission rate. Newspaper, for example, can undergo glowing ignition at a thermal irradiance of 5 cal/cm² from a 100 kT weapon, and at about 7 cal/cm² for a 1 MT weapon. As shown in Table 12, these exposure values are met or

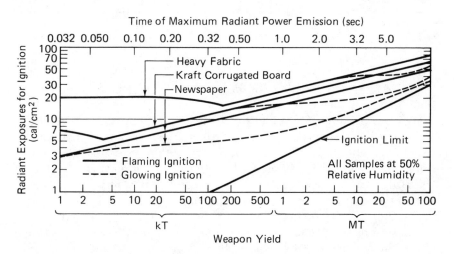

Figure 3. Radiant energy levels necessary for ignition from thermal radiation of nuclear detonations. (From Broido, 1963)

Table 12.

A. *Overpressure—Thermal Radiation Distance Relationships (air bursts)[a]*

Weapon yield (kT)	Peak overpressure (psi)	Ground range (km)	Slant range (km)	Circumference (km)	Q (cal/cm^2)
100	2	5.64	5.85	35.43	5.70
	3	4.35	4.62	27.33	10.44
200	2	6.77	7.04	42.54	7.87
	3	5.64	5.97	35.43	10.94
300	2	8.06	8.36	50.62	8.37
	3	6.44	6.82	40.49	12.57
500	2	9.67	10.02	60.74	9.71
	3	7.57	8.02	49.58	15.15
1000	2	12.41	12.85	77.95	10.12
	3	9.67	10.23	60.74	18.63

B. *Overpressure—Thermal Radiation Time Relationships*

Weapon yield (kT)	Time to 70% thermal irradiance (sec)	Blast wave arrival time at 2 psi[b] (sec)	
		Surface burst	Air burst
100	1.7	8.0	16
200	2.3	10.0	20
300	2.8	11.5	22
500	3.5	13.0	28
1000	4.7	17.0	35

[a]Calculations and data following Glastone and Dolan (1977).

[b]Differences in blast wave arrival times for air vs. surface bursts reflect slant ranges vs. ground ranges.

exceeded at the 2 psi isopleth for the weapon yields in the scenario being examined here.

High probabilities for urban fire ignition are frequently assigned to the 1–3 psi isopleths (DCPA, 1973). Miller (1963) estimated that under visibility conditions of 15 km, kindling fuel within residential structures will ignite out to the 3 psi range, where Q values typically exceed 10 cal/cm^2 (Table 12A). The OTA (1979) report assumed that at the 5 psi range, 10% of the exposed buildings would burn, and at the 2 psi range, 2% of the buildings would be involved. Wiersma et al. (1973) concluded that at the 2 psi isopleth, very few room fires may succeed to "flashover", that is to total involvement of the room in flame, but only a few of these room fires are necessary for initiation of mass urban fires.

The interaction of blast wave effects with fire ignition and spread is poorly understood. It is important to note that the thermal pulse travels at the speed of light, whereas blast waves travel at supersonic or subsonic speeds. For example, under ideal conditions peak thermal irradiance produced by a 300 kT weapon will arrive at a distance equivalent to the 2 psi isopleth almost 9 sec in advance of the blast wave for a surface burst, and 19 sec in advance of the blast wave for an airburst (Table 12B). Conflicting results from various studies indicate that the dynamic overpressure associated with the blast wave can either extinguish fires ignited by the thermal pulse or rearrange fuels to enhance combustion (DCPA, 1973; Martin, 1974; Wiersma et al., 1973). In addition to these phenomena, however, the blast wave ruptures fuel storage tanks and lines, disrupts industrial facilities, exposes new fuels, and otherwise can initiate secondary fires, particularly as ignited by residual flames from the thermal pulse.

Ignition of interior kindling fuels occurs from exposure to the thermal pulse through windows. Hence, directionality relative to the burst and exterior shielding of windows by trees or other structures is important to consider. DCPA (1973) estimated that probability of shielding in residential areas of 1–2 story buildings ranges between 27–34%. Similar values for residential high rise areas of 3–4 story dwellings would be 46–54%, and 88–92% in areas of multistory commercial buildings.

Once ignition of some interior kindling fuels has occurred, flashover is determined by the types of fuels present (e.g., draperies, upholstery, beddings). Flashover under normal conditions may require 30–90 minutes (DCPA, 1973; Martin, 1974). However, glass breakage is severe out to the 0.5–1 psi range (Glasstone and Dolan, 1977), indicating that at that range the dynamic overpressure seen as 35 mph winds would be sufficient substantially to rearrange interior fuels, enhancing the probability and rate of flashover in instances where primary ignitions have not been extinguished. Indeed, if we accept 2 psi as the margin for effective thermal irradiance, extensive fuel redistribution is assured by the degree of damage that buildings at that range would incur (Glasstone and Dolan, 1977).

Further discussion of urban fire spread is complicated by the fact that blast damage to most residential and office structures within 1–3 km of the epicenter would be severe. Hence, the fuel bed to that range would consist of an uneven distribution of nonflammable rubble and flammable debris. Secondary ignitions from ruptured gas lines and other sources would be common in this area. Past experience suggests that within an area of moderate to severe building damage, approximately six secondary fires can be expected per 10^5 m^2 of building floor area; in an area with a building density of 25%, there is a high probability for the occurrence of about 30 building fires per km^2 (DCPA, 1973).

Firespread between buildings occurs via firebrands (flaming or glowing embers carried by the wind), radiant exposure, and/or piloted ignition (also known as convective spread) (DCPA, 1973; Chandler et al., 1963). Building density is an important variable for all of these mechanisms, but especially so for radiant and piloted spread. Piloted ignition is the simplest and most limited mechanism, as direct flame contact is required.

Firebrands can be carried over long distance and ignite fires in areas previously spared. About 82% of 130 major buildings studied from the Hiroshima experience were ultimately burned out, but only 20% of these were burning within 30 minutes of the nuclear explosion, and some were not ignited until 15 hours later, primarily by firebrands (DCPA, 1973). Two concrete reinforced bank buildings survived initial blast/shock and thermal effects of the nuclear explosion, but were threatened by secondary fires ignited by firebrands 1–2 hours later. Survivors seeking shelter in these structures managed to extinguish most of these secondary fires, limiting subsequent damage to single floors.

Radiative fire spread occurs when incident radiant exposure on an ignitable surface, such as wood, exceeds 0.4 cal cm^2 sec^{-1}. Normal flames emit heat energy at a rate of 4 cal cm^2 sec^{-1} (DCPA, 1973; Chandler et al., 1963). Empirically derived probabilities suggest that at a distance of 6 m, there is a 65% chance that fires will spread. Beyond 25 m, radiative fire spread is negligible (DCPA, 1973). Hence, radiative spread is most effective within city blocks; most thoroughfares provide sufficient gaps to preclude block-to-block spread.

The net effect of the preceeding review is to accept the 2 psi isopleth as the range for thermal pulse- and blast-induced fires in urban areas.

Although firestorms and conflagrations are the two mass fire categories most frequently identified (Chandler et al., 1963; Martin, 1974), other authors distinguish group fires as a third category approaching the severity of firestorms (FEMA, 1982b). Conflagrations are distinguished as having a moving fire front, originating from a single ignition point or merged ignitions, with subsequent spread maintained by ambient winds (Broido, 1963; Martin, 1974). Conflagrations continue until a barrier to fire spread is encountered (e.g., surface water) or weather conditions become unfavorable.

Firestorms, and similar but less powerful group fires, occur under specific weather/fuel/fire conditions that are not well understood. From an analysis of WWII data, and an extensive research program carried out in the 1960's, a general concensus has emerged that firestorms require:

greater than 8 lb ft^{-2} of fuel; greater than 50% of the structures in a minimum area of 0.5 mi^2 ignited; and surface winds of less than 8 mph. Subsequent fire spread is contained by convective winds drawn into the area of the fire. Fatalities associated with firestorms result primarily from high temperatures and carbon monoxide (CO) (Bond, 1946; DCPA, 1973), although asphyxiation is sometimes reported as having been insignificant in some cases (Mark, 1976). The duration of mass fires is determined by the quantity and nature of fuels present in the involved area.

The power densities of group fires and firestorms occurring in WWII have been calculated (Figure 4; FEMA, 1982b). In this analysis, Hiroshima and Nagasaki are categorized as group fires with power densities of about 50×10^6 BTU mi^{-2} sec^{-1}, and percentage mortalities of roughly 3 and 0.5 for populations at risk. In contrast, the Hamburg firestorm [in three attacks over the period of 24 July–3 August, 1943 (Bond, 1946; Mullaney, 1946)] released about 700×10^6 BTU mi^{-2} sec^{-1}, resulting in an 18% fatality level for the population at risk. Hamburg was an extreme case having fuel loadings of 32 lb ft^{-2}, whereas

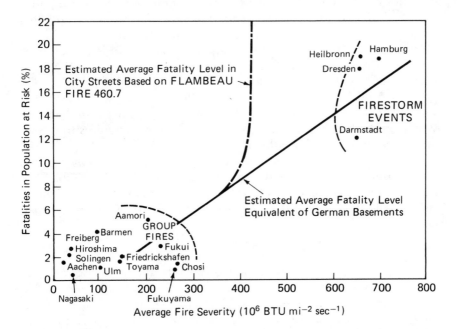

Figure 4. Fatalities as percent of population at risk as a function of fire intensity. (From FEMA, 1982b)

American cities are estimated to have fuel loadings of 2–5 lb ft^{-2} in residential areas (DCPA, 1973; OTA, 1979), with possible values exceeding 50 lb ft^{-2} in commercial and industrial areas.

Example calculations for residential areas reveal that, assuming 8000 BTU released per pound of fuel in an area with a fuel load of 5 lb ft^{-2}, and a fire duration of about 30 min., 56×10^6 BTU mi^{-2} sec^{-1} could be released on the average (DCPA, 1973). With a fuel load of 64 lb ft^{-2} in a densely built up commercial/industrial center, similar calculations reveal a potential average energy release four times that observed in Hamburg.

Fatalities inflicted upon surviving populations, then, have ranged from 0.5–6% in the group fires heretofore observed, and from 12–18% for firestorms. Survivors of these events were in protective shelters or basements. Above a power density of 400×10^6 BTU mi^{-2} sec^{-1}, 100% mortality of unsheltered populations is assured (Figure 4). Assuming a fire duration of 90 min., a fuel load of 9.6 lb ft^{-2} would be sufficient to provide such a power density.

Another complicating factor is that the timing of these urban fires may allow potentially affected people to migrate from the area. Countering that, however, the effects of blast would significantly increase injuries and result in many not being able to move; the blast would destroy the physical structure of the area, so that pathways for exit would be obstructed and tortuous; flash blindness could be widespread, preventing active avoidance of fires during the period of initiation; the local fallout, discussed below, would be greatest during the period when people would be exposed and trying to flee the fires; and those who sought shelters might experience levels of noxious fumes from burning synthetics, high CO levels, and potentially fatal temperatures from the fires surrounding or above them.

Using the 2 psi value as indicative of areas to be consumed by fire, the total populations at risk and casualty figures are listed in Table 13. Fatality rates are assumed to range from 2% to 20% of the population at risk, based on the WWII data. Non-fatal injuries are assumed to equal fatalities.

Initial Ionizing Radiation

When a nuclear detonation occurs, a substantial portion of the energy is emitted as ionizing and electromagnetic radiation. The latter has been dealt with and constitutes by far the largest amount of energy in initial radiation. However, ionizing radiation is so effective in causing bio-

Table 13. Casualties from Fires in Urban Areas.

	Population at risk[a] (air bursts)	Fatalities[b]	Injuries[c]	Total casualties
SMSA cities	1.0×10^7	$2 \times 10^5 – 2 \times 10^6$	$2 \times 10^5 – 2 \times 10^6$	$4 \times 10^5 – 4 \times 10^6$
Largest ten cities	2.3×10^7	$4.5 \times 10^5 – 4.5 \times 10^6$	$4.5 \times 10^5 – 4.5 \times 10^6$	$9 \times 10^5 – 9 \times 10^6$
Total	3.3×10^7	$6.5 \times 10^5 – 6.5 \times 10^6$	$6.5 \times 10^5 – 6.5 \times 10^6$	$1.3 \times 10^6 – 1.3 \times 10^7$

[a]The 2 psi isopleth is taken to include all urban areas consumed by secondary fires. The population at risk is the same as estimated for Table 9D, i.e., those within the 2 psi isopleth minus those within the 5 psi isopleth.

[b]Range of 2-20% of population at risk taken from Figure 4.

[c]Non-fatal injuries assumed equal to fatalities.

logical damage that its relatively low share of energy can cause disproportionately serious consequences. In the following discussion, we briefly overview the nature of initial ionizing radiation and its health effects, largely based on Glasstone and Dolan (1977) and Rotblat (1981).

There are two types of ionizing radiation associated with a nuclear explosion: initial and residual. The initial part is defined to include only that emitted within the first one minute after a detonation. This time period was selected on the basis that the effective range of fission- and fission product-generated gamma rays emitted from a 20 kT warhead is about 3.2 km, so that gamma rays emitted from a source higher above the ground than that would largely be attenuated in the air and not constitute a significant health hazard (Glasstone and Dolan, 1977). The fireball of an air burst rises rapidly in the surrounding, cooler air, reaching the 3.2 km level in about one minute from a 20 kT near-surface air burst. The effect of increasing warhead size and, consequently, the optimal air burst height counteract, so that the one minute period is still approximately correct. That is, much larger warheads emit much greater amounts of gamma rays, resulting in an increased effective range before atmospheric attenuation; however, the burst would likely occur at a greater height, and the resultant cloud would ascend at an increased rate.

Five kinds of radiation are involved in nuclear detonations: alpha particles, beta radiation, gamma rays, X-rays, and neutrons. The alpha particles are either from the fusion process itself or from alpha-emitting heavy radioisotopes associated with fission. The latter is primarily important only as such debris is dispersed and becomes a part of local and global fallout (i.e., residual radiation), a topic to be discussed later. The alpha particles emitted from fusion have very short free path lengths prior to being absorbed in the air; thus initial alpha radiation would only be important at distances very close to the fireball itself, where other physical factors would totally predominate in causing damage.

A similar situation exists for beta radiation, whose sources are fission products (important to residual fallout), the fission process itself, and neutron absorption by certain nuclei. Again, free path lengths are relatively short, so initial beta radiation is not important.

X-rays are emitted from the fireball acting much like a black box radiator. As we have seen, at the temperature of $10^7\,^\circ$C at the fireball surface, these are in the thermal X-ray range, quickly become absorbed by the surrounding atmosphere, and eventually lead to blast and thermal radiation effects.

The remaining two types, neutrons and gamma rays, constitute the important component of initial ionizing radiation.

Almost all (99%) of the neutrons are emitted from the fusion and/or fission processes within 1 microsecond into the detonation; these constitute prompt neutrons. About 1% is emitted later (mostly within the first minute) from continuing fission reactions from weapon debris. The total quantity of prompt neutrons emitted is proportional to the warhead yield, but the specific amount released depends on the specific weapon configuration, with about an order of magnitude more neutrons from fusion than fission. [Enhanced radiation weapons, also known as neutron bombs, do not have the fission-fusion-fission chain of typical thermonuclear bombs, lacking the last fission step by not having an outer shell made of fissile material (see Cohen, 1978). Thus, fusion-generated neutrons are not absorbed by the outer shell but are released in tremendous quantities to the surrounding air.] Fission neutrons are emitted in a continuous spectrum of energies (reflected in velocities of the neutron particles), whereas fusion neutrons are initially limited to a particular energy level associated with deuterium-tritium reactions. However, after a number of collisions with other nuclei, a broader spectrum of neutron energies ensues from fusion sources.

The neutron flux decreases with the square of distance from the source, in a per unit area effect, but neutrons also are attenuated through absorption in the atmosphere. The relationship is exponential, with about 90% absorbed within 600 m, but that value varies substantially with air density. Dose values are a function of the number of neutrons and their kinetic energies. Air bursts give about twice the dose as surface bursts (half the latter neutrons are absorbed by the ground). Rotblat (1981) presents a graph of neutron doses as a function of slant range; interpolations show 100 kT to give a dose of about 400 rads at 1.6 km, and 1 MT at 2.0 km. In general, the dose from neutrons decreases with distance much faster than the effects of blast and thermal radiation.

Gamma rays comprise the other major type of initial ionizing radiation. They are emitted from a number of sources: (1) prompt gamma rays result from the fission process itself, are of highest intensity, but almost all are absorbed within the weapon itself; (2) fast neutrons colliding with nuclei in both the bomb debris and the air can transfer some of their kinetic energy to increase the energy state of the nuclei, which subsequently release the excess as gamma rays (known as inelastic scattering); (3) neutrons may also be absorbed by nuclei, especially slower neutrons that have lost some kinetic energy via scattering; this process, called radiative capture, results in an unstable nucleus that subsequently releases a gamma ray; it is the dominant source of gamma rays at most distances; (4) "delayed" gamma rays are emitted continuously from fission products and constitute a major source of initial and fallout radioactivity. An important phenomenon with delayed gamma rays is the considerable hydrodynamic enhancement of gamma rays at some distance from the detonation after a blast wave has passed, leaving air pressures reduced and, consequently, a substantial reduction in absorption of the gamma rays by the atmosphere. This effect can make a weapon have an effective yield with respect to gamma radiation many times higher than the actual yield (factor of 2–5 increase for 100 kT, factor of 20–50 increase for 1 MT, factor of 10^3–10^4 increase for 10 MT) (Glasstone and Dolan, 1977).

Gamma rays are attenuated in air exponentially, but not quite as readily as neutrons. Thus, at greater distances, gamma rays become more important than neutrons. Air burst gamma doses according to Rotblat (1981) are 400 rads at 1.8 km for 100 kT and at 2.6 km for 1 MT (corrected for hydrodynamic enhancement).

Radiation Health Effects

In this section we discuss the human health implications from ionizing radiation when the exposure is acute and doses are substantial; in a later section (dealing with long-term radioactive fallout), we will discuss chronic effects from lower levels of radiation. First, some terminology requires clarification.

Radiation results in damage to biological systems by the dissipation of the radiation energy through absorption in tissues and the specific transfer of that energy to the creation of ions in the absorbing material. Such ions are reactive chemically, and subsequent chemical transformations can be translated in various ways to adverse biological effects in cells, tissues, and the total organism.

To measure the energy in radiation/matter interactions, the *radiation exposure* is defined as the ability of a given amount of radiation to create ion pairs in air. The *roentgen* (R) is the quantity of gamma or X-rays that produces a total charge of 1 coulomb in 1 kg of air at standard temperature and pressure (2×10^9 ion pairs per cm^3 of air). Many monitoring instruments measure in R.

The value of more importance, however, is the *radiation dose*, the amount of energy of ionizing radiation actually absorbed in tissue. The *rad* is defined as absorption of 100 ergs/g of material; key here is that the rad level depends on the absorbing medium as well as the radiation.

Thus, no simple translation from R to rads exists. For gamma rays in air, 1 R = 0.87 rad, but in water (and, thus, in soft biological tissues), 1 R $\cong$ 1 rad.

To determine biological effects, however, the rad is insufficient alone. Different types of radiation cause different amounts of biological damage, even for the same absorbed energy. The reason is that some radiation disperses its energy very rapidly, being slowed down and stopped within small distances of tissues. In that situation, the energy per unit distance would be quite high, and damage is concentrated. The term *linear energy transfer* (LET) relates to this phenomenon, where high-LET radiation has high biological damage. The ratio of the LET for a particular type of radiation to the LET for low-LET radiation is the *relative biological effectiveness* (RBE). The biological damage is measured as a dose equivalent, which is the dose (in rads) times the RBE. The unit for dose equivalent is *rem*. RBE factors are 1 for beta radiation, gamma rays, and X-rays; 10–20 for alpha particles; and for neutrons, 1 for acute effects and 4–10 for long-term effects. Thus, for example, one rad of alpha dose will cause the same damage as 10 to 20 rads of gamma rays.

In the current case of initial ionizing radiation, only gamma rays and acute neutrons are important, so 1 rad = 1 rem. The effects of acute exposure to gamma rays and fast neutrons are similar and, for purposes here, can be treated as additive. Unlike thermal radiation, shielding is not effective, particularly from gamma rays, which are highly scattered, providing exposure from all directions (termed "skyshine").

The acute dose/response curve for humans has a rather steep slope, so that below 200 rem, no fatalities occur, and above 600 rem, almost 100% fatalities occur. It is generally accepted that the LD_{50} for healthy adult humans provided with appropriate medical care is about 450 rem (Rotblat, 1981; Glasstone and Dolan, 1977; Coggle and Lindop, 1982). Under conditions of an actual nuclear war, and considering effects on the more sensitive young and elderly members of a population, a lower LD_{50} would be more appropriate. Synergisms of radiation dose and other factors will be discussed later. For the purposes here, we will use the acute LD_{50} value of 450 rem, providing a conservative estimate of fatalities. A summary of acute radiation illness is provided in Table 14.

From Glasstone and Dolan (1977) and Rotblat (1981), estimates can be made of the distances to which particular equivalent doses from gamma rays and neutrons are received. Treating these two components as additive, the distances to 450 rem are listed in Table 15. Also shown are the distances for the 5 psi lethal area from blast.

Table 14. Acute Radiation Sickness.

Dose (rem)	Health effect/fatalities[a]
150	No deaths. 50% gastrointestinal (GI) distress (prodromal syndrome).
200–1000	0–100% mortality. Effects from damage to bone marrow (hemopoietic distress); death delayed to 30 days. $LD_{50} = 450$ rem for healthy adults.
1000–5000	100% mortality in 7–14 days; kills epithelial cells of GI tract plus hemopoietic effects.
>5000	100% mortality in 48 hr; central nervous system damage and failure.

[a]Summarized from Glasstone and Dolan (1977) and from Coggle and Lindrop (1982).

The comparison of the latter shows that for all weapons used in the present scenario, those individuals close enough to absorb a lethal acute dose would be well within the blast lethal area. Similarly, all persons exposed to the minimal dose needed for acute radiation sickness would also be within the lethal area from blast. Thus, there are no projected casualties from initial ionizing radiation, since all such victims would already be dead from blast. [This situation did not occur for the bombing of Hiroshima and Nagasaki (Ishikawa and Swain, 1982), since weapons of 10–20 kT were used, for which initial ionizing radiation exposure areas exceeded the blast lethal area.] It should be understood that radiation effects would still be very significant in causing casualties, but this would be limited to radiation from local and global fallout, to be addressed later.

Table 15. Immediate Ionizing Radiation Doses.

Weapon yield (kT)	450 rem slant range[a] (km)	200 rem slant range (km)	Blast lethal area radii 5 psi (km)	
			Air	Surface
100	1.8	2.0	3.3	2.1
200	2.1	2.2	4.1	2.7
300	2.2	2.4	4.7	3.0
500	2.4	2.5	5.6	3.6
1000	2.7	2.8	6.8	4.5

[a]Combined doses from gamma rays and neutrons, summarized from Rotblat (1981) and Glasstone and Dolan (1977).

State of Physical and Biological Systems: Direct Effects

Effects from Blast, Initial Ionizing Radiation

In the previous discussions, we have looked at the major mechanisms for injury and death to humans to occur from the immediate effects of nuclear weapon detonations. These have included the effects of the blast wave, the thermal radiation pulse and associated fires, and the ionizing radiation emitted in the first minute after detonation. To characterize the immediate effects of nuclear explosions on the non-human systems involves the same major mechanisms. However, because targets for the detonations are preferentially located at centers of human activity, these will receive a grossly disproportionate share of the direct effects. This means that human casualties cannot be accurately estimated by just taking the fraction of the land area of the U.S. within the 5 psi peak overpressure isopleth, for example, and applying that fraction to the total U.S. population; the casualties from blast have to be treated on a much more specifically defined basis of area-population relationships. In addition, humans and their systems are often considerably more sensitive to the immediate effects of nuclear weapons than are natural systems. For instance, humans suffer adverse health effects from acute radiation at levels well below most other species. Finally, there is obviously a far greater concern for effects on humans than on other biological systems. Human casualties in the immediate period of 10^7–10^8 dead and injured for the United States alone would be of tremendously more importance than any conceivable deaths of other biota, including complete extinction of species.

Any consideration of the direct effects on physical and biological systems needs to be tempered by this perspective. Nevertheless, it is important to characterize the direct effects on natural systems in order to provide a basis for considering longer term, indirect effects on those systems. That, in turn, is vital to the survivability of those humans who do manage to escape death in the immediate period of a nuclear war.

Most of the direct effects on natural systems from detonations over urban areas can be dispensed with easily. The area covered by blast of 3–4 psi (Table 16) is essentially within the urbanized areas of all of the top ten U.S. cities and for all but the smallest of the city size classes treated statistically in the previous analyses. For the latter, including cities of 100,000 individuals or less, the 3–4 psi radii extend only a short distance beyond the area where the population exceeds the background, average U.S. density (25 individuals/km^2); the area of overlap is rather small and

Table 16. Blast Effects on Forests (Air Bursts).[a]

Peak overpressure (psi)	Peak wind velocities (mph)	Damage (% blowdown)	Warhead yield (kT)	Ground range (km)	Area (km^2)
4	130–140	90	100	3.6	39.6
			200	4.5	63.9
			300	5.0	78.3
			500	6.0	111.6
			1000	7.6	180.0
3	90–100	30	100	4.4	59.4
			200	5.6	99.9
			300	6.4	130.3
			500	7.6	180.0
			1000	9.7	293.8
2	60–80	10	100	5.6	99.9
			200	6.8	144.0
			300	8.1	204.1
			500	9.7	293.8
			1000	12.4	483.8

[a]Calculations and data following Glasstone and Dolan (1977).

not a substantial portion of the area for natural systems. Thus, with respect to blast from urban bombings, effects on natural systems can be ignored.

A similar situation exists for the initial ionizing radiation, where, as we have seen, doses sufficient to cause non-lethal acute illness in humans are only experienced within areas lethal from blast effects. Further, as Woodwell (1982) reported, LD$_{50}$ values for other species are virtually all in excess of dose/response values for humans. Thus, initial radiation effects on plants and other animals would only appear very close to the detonations, where other mechanisms of injury would well dominate.

The effects from the thermal pulse are somewhat more complicated, in that the areas are larger than blast or radiation effects areas and, thus, may extend further into non-urban regions. Of particular importance is the possibility of generating fires that become self-sustaining and consequently could cover much larger areas of natural systems. This problem will be explored below.

In short, the direct effects from urban detonations would not cause an appreciable impact on natural systems, except perhaps for fires. Effects from non-urban detonations, particularly on military targets, do offer the potential for more impacts on natural systems. Again, though, radiation and blast areas are relatively limited. The specific natural systems around most such targets are already relatively expendable, and they do not provide substantial support to humans and society. Therefore, in general direct loss of these areas to blast and radiation effects does not constitute a major portion of natural systems. However, the initiation of fires that spread beyond the initially affected areas can potentially lead to large-scale destruction of certain types of eco-systems.

Fire Initiation

The incendiary potential of nuclear weapons has long been a concern to defense planners. The issue was given stronger emphasis following Congressional hearings with regard to the feasibility of federally sponsored bomb shelter programs and civil defense planning (Strope and Christian, 1964). In addition, the Soviet test detonation of a weapon up to 60 MT in yield strengthened the concern as to how fire aspects of nuclear weapons should be treated. As a result of the Soviet test and poor targeting precision of the arsenals extant in the early 1960's, most analyses focused on weapons in yield ranges of 1 MT–100 MT.

In an analysis of wildfire spread following nuclear war, Ayres (1965) concluded that, depending on targeting and seasonally related climatic conditions, 19,000 mi^2 to 160,000 mi^2 (50,000 km^2 to 410,000 km^2) of forests and rangelands could be consumed. Although Ayres provided an adequate discussion of the factors determining ignition, including kindling fuel type, weapon yield, and climatological factors, this projection is based on the assumption that primary fires from nuclear war would consume from between one-half to two times the total areas involved in wildfires using minimum and maximum record years between 1926–1959. Thus, no direct calculations were attempted in this analysis.

Although several studies have been conducted to analyze specific aspects of fires and nuclear war (e.g., Chandler et al., 1963; Wiersma et al., 1973; Martin, 1974), these have largely dealt with urban fire potential, and no reliable estimates of the potential for natural system fires or their ecological implications have appeared in the literature. In particular, major studies of the consequences of nuclear war have largely overlooked the wildland fire problem. The OTA report (1979) takes a decidedly medical and socio-economic approach in their analyses of various scenarios, presumably lumping wildfire effects with other "incalculable" effects that were considered to be "potentially as serious as those consequences that were calculated." The NAS report (1975) similarly avoided direct consideration of urban as well as wildfires by focusing on longer term effects considered to have global implications for survival (e.g., ozone depletion and UV light). Crutzen and Birks (1982) offered as a working hypothesis that 10^6 km^2 of forests would burn in the Northern Hemisphere following the nuclear war depicted by the Ambio Advisory Group (1982) scenario.

In order to evaluate the areal extent of natural systems subject to fires, the same calculations are used as previously for radiant energy as a function of weapon yield and height, atmospheric transmissivity, and distance along a slant range. Translating this to wildfires requires a look at the requirements, in terms of the radiant exposure (Q values), for effective ignition of materials.

According to Anderson (1970), ignition can be considered as that point in the time-temperature history of a material when either a flame or glowing ember appears. More precisely, ignition requires a complex set of dehydration and pyrolysis reactions that result in the removal of excess moisture from the fuel; it also requires the breakdown of cellulosic constituents to become volatile gases, which pass to the surface of the material and undergo combustion (i.e., rapid oxidation). Three stages can be identified in fire ignition: pre-ignition, flaming phase, and glowing phase.

Among the factors that determine the ignition threshold of a substance are: (1) thermal absorptivity; (2) density of the fuel; (3) fuel specific heat; (4) fuel thermal diffusivity; (5) moisture content; (6) inorganic constituents; (7) radiant exposure/unit area; and (8) duration of radiant exposure (Martin, 1974). The ignition requirements of various fuels can be calculated using an empirically derived formulation of the above factors; the predictability of the available formulas has proven to be acceptable in various weapons tests (Martin, 1974; see also, Anderson, 1970).

Glasstone and Dolan (1977) presented data on fuel ignition thresholds from several nuclear weapon tests. The importance of fuel color (i.e., as it affects absorptivity) is clear from these data. The data in Glasstone and Dolan (1977) and the curves presented in Figure 3 further illustrate the importance of weapon yield to ignition. From the information available on natural fuel ignition characteristics, and assuming low moisture contents, the values in Table 17 are taken to be the representative ignition thresholds for the listed fuels.

These values are met or exceeded for most materials at the 2 psi isopleth for all weapon yields that we are considering (Table 12). Strope and Christian (1964) have similarly suggested that a probable ignition radius would extend to between the 1–3 psi isopleths. Ayres (1965)

Table 17. Representative Ignition Thresholds.[a]

Fuel	Radiant exposure Q (cal/cm^2)
Rotted wood	5.0
Deciduous leaves	5.0
Fine grass	6.5
Coarse grass	7.5
Pine needles	13.5

[a]Information derived from Glasstone and Dolan (1977).

arrived at a higher ignition threshold of 10 cal/cm^2 (also the level used in Turco et al., 1983a) because of the higher weapon yields he was concerned with; however, even this radiant exposure level is exceeded at the 3 psi isopleth for most of the yields in this study.

There are several additional factors that affect the incendiary potential of nuclear weapons. Among these are shielding of kindling fuels, angle of radiant energy incidence, and blast wave-ignition interactions.

As previously discussed, atmospheric attenuation has a substantial impact on surface radiant expsosure. Figure 5 depicts the average annual cloudiness across the U.S. as a percentage of days when cloud

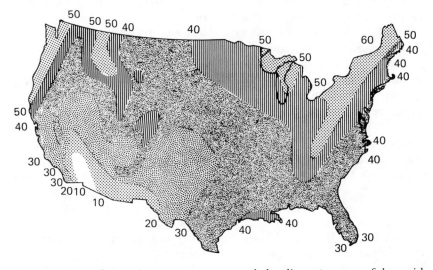

Figure 5. Map of United States average annual cloudiness (percent of days with attenuation ≥ 0.1). (From FEMA, 1982b)

opaqueness provides significant attenuation. In addition, weather data summarized through 1964 (Strope and Christian, 1964) revealed that the average U.S. city has about 130 days of heavy cloud or dense fog per year and rain for about 110 days per year.

Natural wildfires generally begin and are sustained by a fuel bed of dead foliage and branch litter that accumulates on the ground surface over a period of years. In grassland and shrub systems, the fuel bed is more exposed to incident radiations than in forests because of lack of shielding by foliage and tree stems. In addition, shielding of the forest floor increases for a given area as the angle of incidence for incoming radiation increases. Figure 6 illustrates the relationship between angle of incidence and probability of exposure for forest floors with different densities of trees. These data show that line-of-sight elevation angle for an airburst would have to exceed 45° for the probability of radiant exposure to approach 80% in a forest of low density (52 trees/acre).

Calculations for the air bursts where the height equals the ground range (i.e., at a 45° angle) indicate Q values of 50–100 cal/cm^2 for 100–1000 kT weapons. However, these distances are associated with peak overpressures of $\geq$15 psi, with dynamic pressures of 350 mph winds, conditions where the forest would be otherwise destroyed by blast effects and the effective shielding of the forest floor would be moot.

This example highlights the potential interaction between blast wave, fuel ignition, and fuel bed configuration. Depending on terrain, windspeeds approach 160 mph at the 5 psi isopleth, and about 70 mph at the 2 psi isopleth (Glasstone and Dolan, 1977). Although the controlling factors are not well understood, windspeeds of these magnitudes could extinguish some fires, while facilitating the spread of others. Boundary layer phenomena probably determine whether ignitions will be extinguished (i.e., by cooling and removal of pyrolytic gases), whereas improved oxygen availability and alterations in fuel exposure for ignition probably facilitates firespread. Wiersma et al. (1973) found that in one experimental situation using kerosene fuels, windspeeds associated with peak overpressures of 5 psi and lower did not extinguish pre-set fires. However, field reports of grass fires being extinguished at windspeeds exceeding 60 mph can be found (T. Quinn, U.S. Forest Service, Albuquerque, NM, personal communication).

The potential for physical destruction of forests is extreme above windspeeds of 130 mph (see Table 16). The importance of canopy destruction to fire spread is difficult to predict but would probably depend upon species composition, meteorological conditions, and tree density. It is important to note that the blast wave travels at supersonic and subsonic speeds, whereas radiant energy travels at the speed of light.

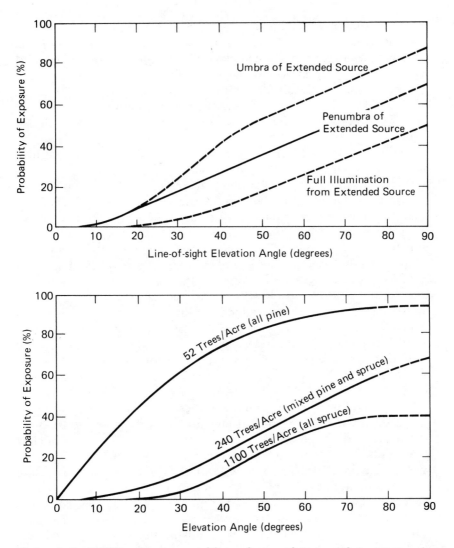

Figure 6. Probability of exposure of forest floor as function of elevation angle and tree density. (From Martin, 1974)

Thus, the peak emissions of radiant energy that occur within a few seconds of burst arrive well in advance of the blast wave, at sufficient times for many kindling fuels to ignite.

In order for wildfires to become established, several conditions regarding fuel bed density and configuration must be met under moisture conditions that are uncommon to rare, or at least seasonally so, in many parts of the U.S. Table 18 summarizes prescribed fire guidelines for several ecosystem types. The data presented reflect the minimal or least favorable conditions under which fires can be established by artificial means and can be sustained long enough to fulfill some management objective. Thus, we can assume that for wildfires, meteorological conditions and fuel loads would have to exceed the tabular values.

The issue thus arises as to how frequently the conditions specified above are met or exceeded in the U.S. Chandler et al. (1963) provide an analysis of this problem, as shown in Tables 19 and 20. The "no spread" criteria were found to be conservative in that 100% of all ignitions meeting these conditions failed to spread, whereas 60% of fire ignitions studied that met "will spread" criteria failed to do so. The regional distributions of "no spread" conditions depicted in the preceding tables clearly reflect seasonal climatological patterns that are further illustrated in Figure 7. Here we see that the most densely populated areas of the U.S., i.e., those areas most likely to be primary targets, have the shortest fire seasons in normal years. Hidden in the analysis is that those areas where fires are most likely (e.g., SW regions) also have the greatest heterogeneity in fuel type and cover.

In order to assess the possible expanse of wildland fires in the United States following a nuclear war, a scenario involving specified military targets was analyzed with specific attention given to vegetation types at risk, wildland fire potential, and the properties of thermal irradiance, as previously discussed. The military targets identified in this scenario were selected for their probable strategic and defense roles as nuclear weapons facilities, following designations in the Ambio Advisory Group (1982) scenario. ICBM silos were presumed to receive 2–500 kT surface bursts each; strategic air bases, air defense facilities, certain naval air stations and warship ports, nuclear weapons storage facilities, and other target categories were assumed to receive 1–300 kT surface burst each.

Two models were used to estimate wildland fire expanses, one based on firespread from an initial circular area of involvement, the other based on wedge-shaped firespread downwind from single ignition points (Ayres, 1965). For each of these models, calculations were made for firespread under "no spread", "actionable", and "critical" conditions (Chandler et al., 1963). No spread conditions were assumed to result in fires only within the 2 psi isopleth; actionable conditions were assumed to result in an ignition point every 500 m along the perimeter on the leeward margin of the 2 psi isopleth; critical conditions similarly were assumed to have an ignition point every 100 m. Data from these

Table 18. Minimal Conditions for Fire Management.[a]

System	Fuel load (kg/ha)	Fuel moisture (%)	Rel. hum. (%)	Air Temperature (°C)	wind (km/hr)	Season
Semidesert grass-shrub	670	20	15–40	21–27	13–24	Spring
Shortgrass prairie	2000	20	30–50	16–24	8–16	Feb–March
Mixed grass prairie	3400		25–40	21–27	13–24	Feb–March
Tallgrass prairie	3400–4500	20	40–60	16–24	8–16	Feb–March
Fescue prairie	1100	20	40–60	4–24	3–19	Feb–March
Marshes	7100	20	30–40	10–18	0–3	Early Spring
Pinyon-juniper woodlands	670	20	20–40	21–23	16–32	Spring
Big sagebrush-grassland	670	20	15–20	24–29	13–24	Early Spring or Fall
Pines—S.E. U.S.	18,000	20–25	30–50	7–10	8–29	Fall
Ponderosa pine	Heavy	15	20–40	4–16	8–24	Fall
	Light	8–12	20	24	24	Fall
Douglas fir	15,000–20,000	9–18	40–55	13–24	3–6	Early Spring
Aspen live	N.A.	20	15–30	18–27	6–24	Early Spring
dead	N.A.	20	35–50	4–24	3–19	Early Spring
California chaparral			25–40	20–30	16–20	Late Spring
Arizona chaparral			10–30	20–35	6–15	Late Spring

[a]Information summarized from Wright and Bailey (1982). These values represent the least favorable conditions under which fires could be artificially initiated and sustained; wildfires are assumed to require more favorable conditions.

Table 19. Minimal Conditions for Prevention of Fire Spreading.[a]

Fuel type	No spread[b]	Fire out[c]
All fuels	1 in. snow on ground	No spread
Grass	R.H. > 90%[d]	No spread or measurable precipitation
Brush hardwoods	If wind 0–3 mph, then R.H. > 50% If wind 4–10 mph, then R.H. > 75% If wind 11–25 mph, then R.H. > 85%	No spread for 36 hr or 0.1 in. precipitation
Conifer timber	a. 0.25 in. precipitation within 24 hr and: If wind 0–3 mph, then R.H. > 50% If wind 4–10 mph, then R.H. > 75% If wind 11–25 mph, then R.H. > 85%	a. 0.5 in. precipitation
	b. 0.25 in. precipitation within 2 days wind as above with respective R.H. 60, 80, 90%	b. No spread for 24 hr and 0.5–0.25 in. precipitation
	c. 0.25 in. precipitation within 4 days and wind 0–3 mph, R.H. > 80%	c. No spread for 96 hr and measurable precipitation
	d. 0.25 in. precipitation within 6 days and wind 0–3 mph, R.H. > 90%.	d. No spread for 7 days

[a]Summarized from Chandler et al. (1963).
[b]No spread = conditions under which fires will not move outward.
[c]Fire out = conditions under which fires will go out without human intervention.
[d]R.H. = relative humidity.

calculations are shown in Table 21. A detailed analysis of the relative risk of different vegetation types using this approach has not been completed, but it is assumed that by comparing no spread, actionable, and critical fire estimates, reasonable projections of wildland fire expanse can be obtained.

It is clear from the analyses that the concentration of weapon strikes on ICBM fields in the Northern Great Plains represents the greatest risk to a vegetation type (specifically, tall-grass and mixed-grass prairie), with projected ranges of fire involvement from 1.9×10^5 km^2 to 2.7×10^5 km^2. It is not known what importance agricultural mosaics would have in promoting or suppressing firespread in some of the ICBM areas, but it is apparent that such interactions would be seasonally dependent.

The coastal sage-chaparral vegetation zones in Southern California are also at great risk, not only as a function of the potential number of targets in that area, but also because of the volatility of component species during dry seasons (i.e., July–October). Also, the number of no spread days during the normal fire season is usually less than 5 per month (Table 20). Firespread in eastern deciduous forests would be highly dependent on the weather history prior to an attack because of the

Table 20. Number of "No Spread" Days at Selected Weather Stations, by Months.[a]

Station	Jan.	Feb.	Mar.	April	May	June	July	Aug.	Sept.	Oct.	Nov.	Dec.
Northern:												
Olympia, Wash.	—	—	24	16	15	16	9	8	15	26	28	31
Boise, Idaho	—	—	—	7	7	3	0	0	1	4	—	—
Casper, Wyoming	—	—	—	9	7	4	2	1	2	4	—	—
Minneapolis, Minn.	—	—	—	9	9	9	9	10	9	11	—	—
Grand Rapids, Mich.	—	—	—	12	10	9	9	10	8	13	—	—
Albany, N.Y.	—	—	—	13	13	11	10	11	14	15	—	—
Washington, D.C.	—	—	12	10	10	8	9	11	10	12	14	—
Central:												
Oakland, Calif.	28	20	13	10	6	3	1	2	2	6	13	22
Cedar City, Utah	—	—	8	3	3	1	1	1	1	3	7	—
Springfield, Mo.	—	—	14	10	10	10	10	7	7	10	12	17
Charleston, W. Va.	—	—	12	12	12	13	12	16	10	14	17	23
Southern:												
Los Angeles, Calif.	16	13	12	13	9	5	1	2	3	6	8	12
Roswell, N. Mexico	4	4	2	1	1	0	1	1	1	3	2	4
San Antonio, Texas	11	10	7	7	8	4	2	2	5	8	10	11
Shreveport, La.	16	13	12	10	10	8	9	7	7	8	12	14
Memphis, Tenn.	21	15	13	9	9	9	10	8	8	9	11	18
Columbia, S.C.	16	13	12	8	7	7	10	11	11	10	11	17
Tallahassee, Fla.	16	12	11	10	11	15	22	19	17	13	14	18

[a]Summarized from Chandler et al. (1963).

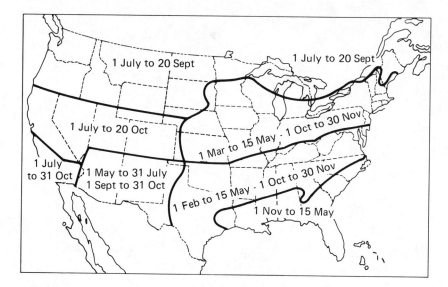

Figure 7. Fire season map of the United States. (From Broido, 1963)

shielding effectiveness of the component species. However, there could be a higher density of urban and industrial targets in this vegetation zone which would increase the probability of fire initiation, even if firespread was limited by unfavorable fuel conditions.

The range of values calculated for the scenario depicted here is 1.9×10^5 km^2 to 3.2×10^5 km^2. By reviewing historical fire records, Ayres (1965) projected a possible range of 5×10^4 km^2 to 4×10^5 km^2 of fire from an unspecified nuclear war. Other studies project values in the lower part of this range (NAS, in prep.).

It is very important to note that a limited scenario involving only military targets has been used in this particular analysis; the seasonally dependent mixture of no spread, actionable, and critical conditions that would be encountered during a combined countervalue-counterforce nuclear attack (including urban areas) could be expected to involve larger areas in wildland fires than projected here.

Other Physical Disturbances

As we have seen, the primary direct effect of nuclear war on natural systems is via the potential for the initiation and spreading of fires; the

Table 21. Potential Wildland Fires Expanse from a Limited Counterforce Attack.[a]

	Areas (km^2)		
	No spread	Actionable	Critical
ICBM sites	1.9×10^5	1.9×10^5	$1.9 \times 10^5 - 2.7 \times 10^5$
Other sites	5×10^3	5×10^3	$7.7 \times 10^3 - 4.7 \times 10^4$
Total	1.9×10^5	1.9×10^5	$1.9 \times 10^5 - 3.2 \times 10^5$

[a]Includes only detonations over U.S. military bases that are not located in urban areas. Two models were used, giving the range of values. Model 1 involves firespread from initial circular area; Model 2 involves wedge-shape firespread downwind from single ignition points.

effects from radiation and blast are considered to be of relatively lesser importance.

Other immediate, but indirect, effects on natural systems may be envisioned. The intent here is only to identify these effects rather than provide a substantive analysis.

One potentially consequential indirect effect relates to the detonation of nuclear weapons at major water control facilities. Especially important could be bursts at the major dams of the western waterways. Of the forty or so major dams in the United States, defined as having a gross capacity in excess of 10^5 m^3 in the reservoir, the largest numbers occur in Arizona, California, Montana, and Washington (U.S. Department of Interior, 1980). California, for example, has nine such dams, with capacities ranging from 3×10^5 m^3 to 6×10^6 m^3. There are six dams (two in Arizona, one each in Montana, South Dakota, and Washington), with capacities $1-4 \times 10^7$ m^3. Sudden release of stored waters could result in widespread flooding, putting at risk millions of people directly (e.g., the city of Portland, Oregon, if part or all of the Columbia River dam system were to fail) or indirectly, such as by inhibiting escape routes from targeted areas, or affecting food and water supplies. In addition, regional-scale effects on natural systems are possible in some areas as the landscape became scoured away.

Another potential effect is for nuclear detonations to induce landslides or avalanches. These would be highly dependent on time of year and recent weather conditions (related to the saturation of the ground with water, the amount of snowmelt underway, etc.). In no case, however, would these effects extend beyond local levels.

A possible anti-submarine warfare (ASW) strategy is the barrage tactic of detonating very large warheads (1–10 MT yield each) in a continuous string of underwater bursts along coastal regions. The intent would be to incapacitate strategic or fast-attack nuclear submarines without knowing where any single specific sub was located, but relying on statistical information concerning probable areas of submarines. The potential physical disruption of this scenario could be extensive in terms of physical damage to habitats, direct mortality of fish and other food resource species, release of substantial quantities of radioactivity to the water column (though not to the atmosphere), and temporary increases in sediment loading on a wide scale. Such effects could be important largely in the context of affecting long-term utilization of those areas for food production for surviving humans.

A similar disproportional effect could be expected on coastal and estuarine ecosystems in general, since urban areas in the United States are largely located in coastal regions. Thus, effects from blast, thermal radiation, and local fallout would be much greater on estuaries and coastal zones than on natural ecosystems in general. Habitat destruction could be so extensive as to affect a significant fraction of these types of ecosystems.

State of Atmospheric Systems

As discussed previously, the present analyses are linked to an evaluation of the global atmospheric consequences of nuclear war by Turco et al. (1983a, b). Their new emphasis was on the results of heavy loadings into the atmosphere of particulates from fires induced by a large-scale nuclear war, including in particular substantial alterations in climatic conditions. The analyses presented in this report follow in large part from the assumption that the climatic conditions projected by Turco et al. (1983a) do reflect the situation after a nuclear war. It should be noted that a series of independent studies have been or are being conducted to confirm or refute the analyses of Turco et al., including Crutzen and Galbally (1984), Covey et al. (1984), MacCracken (1983), a major effort by Soviet scientists (Aleksandrov and Stenchikov, 1983), as well as new studies by the U.S. National Academy of Sciences, the Royal Society of Canada, and the SCOPE Committee of the International Council of Scientific Unions. As of this writing, the climatic projections of a nuclear winter from a large-scale nuclear war are consistently supported by the recent studies. It is useful, then, to recapitulate here the essence of the atmospheric projections, based on the cited reports and the technical support document (Turco et al., 1983b).

The issue of the consequences of atmospheric loadings from nuclear war relates to the hypothesis that already in the history of the Earth dense clouds of particulates may have initiated mass extinctions of species. Initial analysis of the hypothesis that immense quantities of sooty smoke could be released from nuclear war-induced fires and strongly attenuate sunlight was done by Crutzen and Birks (1982). Using new data and refined models, Turco et al. (1983a,b) calculated the climatic effects of dust and smoke clouds. They concluded that after a major nuclear war, thousands of such clouds could blanket the Northern Hemisphere at mid-latitudes in a period of days to weeks after the war. This could substantially alter the radiative balance of sunlight interactions with the atmosphere and the Earth's surface in such a way that perturbations in the atmospheric circulation patterns and associated climate could occur on a hemispheric scale. Further, increased atmospheric circulation between hemispheres, with more rapid transport of particulates in the troposphere and stratosphere from the Northern to the Southern Hemisphere, a phenomenon ignored in previous evaluations, could involve the entire planet in significant climatic alterations.

To study these effects, the authors used a series of physical models, calculating the smoke, dust, radioactivity, and NO_x injections as a function of altitude for each type of nuclear detonation and summing across the number of such detonations in a variety of hypothetical war scenarios. Other models predicted the temporal changes in horizontal distribution of the dust and smoke clouds. Finally, another model dealt with the radiative-convective relationships of particulates and the optical and temperature patterns of the atmosphere as a function of altitude. The scenarios included a 5000 MT nuclear war [which closely relates to the scenarios used in this analysis and in the Ambio Advisory Group (1982) scenario], a 10,000 MT war (for a look at the effects of scale), and a few hundred MT war (to investigate the robustness of conclusions with respect to less extensive scenarios).

Their results suggest that initial fires would be confined to 30°–60°N latitude, with zonal spreading preferentially along latitudinal lines, resulting in the greatest effects at those latitudes but also transporting effects elsewhere. Urban fires would cause the greatest problems, both because of the preferential targeting of urban areas and the increased likelihood of urban fires injecting particulates into the upper atmosphere. After one month, optical depths would be about 4, largely resulting from contributions by tropospheric smoke and secondarily by stratospheric dust. (Optical depth is the exponent of an attenuation factor.) Beyond 2–3 months, the stratospheric dust predominates, as the lower altitude soot is deposited on the surface by rain and dry deposition. Estimates are that the total smoke emission would be 5×10^8 tons released over a few days. Importantly, parametric analyses indicated that scenarios in the 1000–10,000 MT range could generate qualitatively the same effects, with comparable optical depths and associated climatic alterations. Even the most reduced nuclear war scenario analyzed, 100 MT, resulted in large smoke opacities over the hemisphere, suggesting a low threshold for major optical and climatic consequences of nuclear war.

Figure 8 shows the Northern Hemisphere, mid-latitude continental near-surface air temperature deviations calculated from the dust and smoke optical depths of the various scenarios. Particularly important is

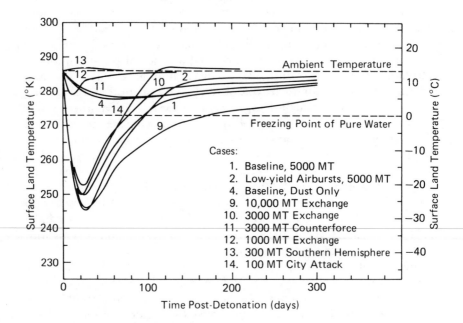

Figure 8. Average air temperatures for the Northern Hemisphere from nine scenarios of nuclear war averaged over diurnal cycles and over the hemisphere. (From Turco et al., 1983a; copyright American Association for the Advancement of Science, 1983)

the tremendous decrease in temperatures within a 3–4 week period after the nuclear war, with a minimum of $-23°C$ experienced 3 weeks after the base case (5000 MT) war. Subfreezing temperatures would endure for several months; even for a 100 MT war, that condition would exist for two months.

Sensitivity analyses indicated that tropospheric soot layers would result in the sudden cooling of the near-surface air that would last a relatively short time, and that stratospheric dust would result in prolonged cooling in periods exceeding the first year. Secondary climatic effects, mediated by alterations in ocean temperatures and current patterns (though marine surface water temperatures would not change much), are possible, enhancing the effects in the projections (see, e.g., Robock, 1984). Clearly, however, air temperatures in maritime regions would not be nearly so reduced as in mid-continental areas, raising the likelihood of extreme weather conditions as air masses would be driven by large differential thermal patterns. Global atmospheric circulation could be expected to experience major alterations, suggesting increased communication of air masses across the equator; predictions of such effects were not included in the Turco et al. (1983a,b) study, but have been reported by Covey et al. (1984).

Changes in surface insolation were also analyzed (see Figure 9). The base scenario resulted in an order-of-magnitude reduction in solar insolation for several weeks, and insolation could fall below the compensation point for months from various scenarios.

Other projections are for decreased precipitation over continental areas for long periods of time, but the temperature gradients between continental and maritime areas could result in heavy precipitation and dense fog in the interface areas. Long-term climatic changes, as from CO_2 injections and NO_x-O_3 interactions, were not considered to be significant. However, O_3 reductions could lead to increases in ultraviolet radiation (UV-B) at the surface, once the particulates cleared from the atmosphere, at levels comparable to previous studies (Crutzen and Birks, 1982; NAS, 1975).

The Turco et al. (1983a, b) study provides the first rigorous analyses that show there would be massive, global perturbations in the temperature and light regimes, particularly in mid-latitude continental areas. The sudden onset of an unprecedented nuclear winter could produce arctic-like conditions in semi- tropical regions. The consequential effects on humans and other biologic systems would be so intense that much of the uncertainty in relating human/biotic/social systems to physical parameters would be lost. The conditions are so extreme that the indirect

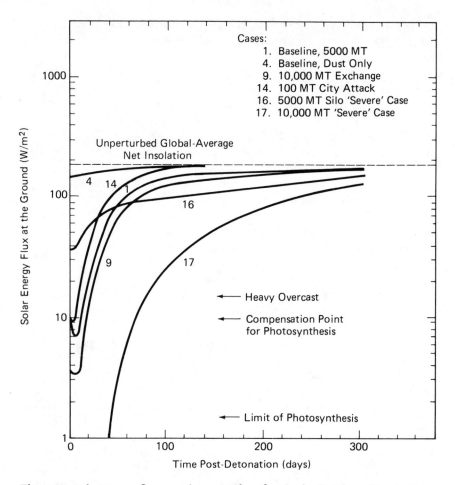

Figure 9. Solar energy fluxes at the ground surface in the Northern Hemisphere from nuclear war, averaged over diurnal cycles and over the hemisphere. (From Turco et al., 1983a; copyright American Association for the Advancement of Science, 1983)

effects they induce could exceed the direct consequences previously documented here. A very important aspect of these findings is that qualitatively the same phenomena would ensue from relatively modest nuclear wars, down to 1000 MT or below. This robustness of the projections in the face of uncertain scenarios adds considerable credence to the supposition that a nuclear war of any but the most contained scale has the potential for devastating effects, exported far from the regions covered by direct detonations.

Local Fallout

The radioactivity associated with a nuclear detonation consists of the initial ionizing radiation, released within the first minute after a detonation, and residual radioactivity, which includes all after the one minute period. Residual radiation is categorized into local fallout, that which reaches the ground within the first one- or few-day period, and global or delayed fallout, which enters atmospheric circulation patterns and can take months or even years before deposition on the surface.

All nuclear explosions lead to fallout, but air bursts (which do not contact the ground) have greatly reduced fallout compared to surface bursts, which entrain large quantities of surface materials into the fireball and render them radioactive. For air bursts, there is essentially no local fallout; rather, the residual radioactivity consists of fission product gases that usually enter the stratosphere. Gradually these return to the surface by coagulation and condensation into particulates. The particles from surface bursts, however, are much larger and coalesce more rapidly, so that about half of the total radioactivity is deposited as local fallout (Rotblat, 1981). This is highly dependent on local weather, especially wind velocities and precipitation. The latter can lead to washout of fallout in more limited areas, subjecting them to increased radiation exposures.

Most fallout consists of fission products, i.e., the pair of new nuclei (occasionally three nuclei in ternary fission) that result from the splitting of each atom of uranium or plutonium. Most of these fission products are radioactive, decay by beta emissions (sometimes also gamma rays), and become other nuclides that also are radioactive. In total there are some 300 radionuclides for 36 elements, with average decay chains of length four (Rotblat, 1981). Their individual half-lives vary from very small fractions of seconds to millions of years. Decay is measured in disintegrations per time [1 Curie (Ci) $= 3.7 \times 10^{10}$ Becquerel (Bq) (disintegrations per second)]. Thus, an inverse relationship exists between the activity of a radionuclide (Ci/g) and its half-life. Because of this, the initial activities of fallout are extremely high, but drop rapidly thereafter.

Calculations of the initial amount of each fission product produced can be made by considering that 1 kT produces about 57g of fission products. Table 22 lists the percentage associated with each major fission product. Thus, as an example, for Cesium 137 (^{137}Cs):

$$\begin{array}{c}\text{fission} \\ \text{product} \\ \text{per kT}\end{array} \times \begin{array}{c}\text{fraction} \\ \text{for a} \\ \text{nuclide}\end{array} \times \frac{\text{atoms/mole}}{\text{g/mole}} \times \frac{\frac{1}{2}\,\text{atoms}}{\frac{1}{2}\,\text{life}} \times \text{Ci/B}_q = \text{Ci}$$

$$57\text{g} \times 0.062 \times \frac{6.01 \times 10^{23}}{137} \times \frac{.5}{9.5 \times 10^8 \text{ sec}} \times 2.7 \times 10^{-11} = 220 \text{ Ci}$$

or a total of 220 Ci of ^{137}Cs is produced per kT of fission energy.

Another major source of fallout is from neutron activation of previously stable material. For example, nitrogen in the air is converted to ^{14}C, a biologically important radionuclide. The most prevalent neutron-activation product from weapon debris is ^{55}Fe, producing about 10^2 Ci per kT (Rotblat, 1981). Neutron activation of the ground also occurs, but this is in areas so close to the fireball that other effects predominate. However, neutron-activated residuals can lead to intense doses to individuals entering the near-field blast zone in the period soon

Table 22. Major Fission Products.

Fission product	Fission yield[a] (%)	Radiation type[b]	Half-life[b]
^{140}Barium	5.7	Beta, gamma, X-ray	12.8 days
^{144}Cerium	4.9	Beta, gamma, X-ray	284.3 days
^{134}Cesium	6.6	Beta, gamma, X-ray	2.06 yr
^{135}Cesium	6.0	Beta	2.3×10^6 yr
^{137}Cesium	6.2	Beta	30.2 yr
^{129}Iodine	0.9	Beta, gamma, X-ray	1.6×10^7 yr
^{131}Iodine	3.2	Beta, gamma, X-ray	8.04 days
^{147}Promethium	2.4	Beta	2.62 yr
^{103}Ruthenium	6.6	Beta, gamma, X-ray	39.4 days
^{106}Ruthenium	2.7	Beta	368 days
^{89}Strontium	2.9	Beta	50.6 days
^{90}Strontium	3.2	Beta	28.6 yr
^{99}Technetium	6.3	Beta	2.1×10^5 yr
^{133}Xenon	5.5	Beta, gamma, X-ray	5.25 days
^{91}Yttrium	5.8	Beta	58.5 days
^{93}Zirconium	6.4	Beta	1.5×10^6 yr
^{95}Zirconium	6.3	Beta, gamma	64.0 days

[a]Data compiled from Rotblat (1981).
[b]Data compiled from Kocher (1981).

after detonations, a situation encountered in Hiroshima and Nagasaki (Ishikawa and Swain, 1981). Particularly important are ^{28}Al, ^{56}Mn, and ^{24}Na in ground and building materials, and ^{38}Cl and ^{56}Mn in coastal areas (Rotblat, 1981); each are beta emitters, with half-lives of 2 minutes to 15 hours.

The other component of fallout includes residuals from the weapons themselves, in particular ^{235}U, ^{238}U, and ^{239}Pu from incomplete fission reactions and ^{3}H from incomplete fusion reactions. ^{239}Pu is at low levels of activity, about 3 Ci per kT, but remains very hazardous for long periods of time through alpha emissions (half-life of 2.4×10^4 yr). Tritium can be biologically important by direct incorporation into hydrologic cycles, but its weak beta radiation is not nearly as great a health hazard.

The mechanisms by which this suite of radionuclides in residual radiation can lead to health effects are by: (1) external radiation from a radioactive cloud passing by; (2) internal doses from inhalation of such a cloud; (3) external radiation by gamma rays emitted from radioactive substances deposited or induced on the ground; (4) internal doses from ingestion of the latter via food chains. The third is the major component of local and global fallout doses; the fourth becomes more important after longer periods of time. Thus, in large part the doses from gamma rays in fallout constitute the major health hazard. For our purposes here, that component is primarily emphasized.

Calculations of gamma doses from fallout must take into account the effects of radioactive decay, where each isotope decays exponentially with time. But such a complex mixture of radionuclides as in fallout has individual half-lives varying over tens of orders of magnitudes. Thus, the combined decay rate is a mixture of a large number of exponential functions. This is further complicated by the creation of daughter products that themselves are radioactive. Thus, some radionuclides actually increase over time, as their parent nuclides decay, and then decay at their own rates. Empirical evidence shows that during the first one-half year the mixture decays by a factor of ten for every seven-fold increase in time (e.g., one-tenth the radioactivity at one week than at one day). Beyond one-half year, the decay rate is faster, but not easily characterized (Rotblat, 1981).

Other factors important to calculating dose include the phenomenon of fractionation, where differential deposition of radionuclides occurs (e.g., ^{90}Kr, a noble gas, does not deposit as rapidly as, say, ^{3}H). Also important is the proclivity for resuspension of particulates once deposited (aeolian transport), the shift in the quality of emissions (e.g., gamma and beta decay predominates at first, alpha predominates later), and the possibility of "salting" of the weapon itself to create excessive amounts of fallout. The latter is potentially feasible by adding quantities of stable cobalt to the warhead, producing ^{60}Co by neutron absorption, which is a biologically important source of external radiation from its high energy gamma ray and half-life of over five years. Another complicating factor is that the heavy particulate loadings to the atmosphere projected to occur from large-scale urban and natural fires could affect the rates at which various radionuclides coalesce into particulates of sufficient size to deposit as fallout. In this situation, local fallout levels could be significantly increased over predictions made from empirical studies of individual weapons detonations occurring in the absence of secondary fires.

The relationships have been developed between warhead yield, wind patterns, detonation altitude, and warhead type with the patterns of local fallout. Glasstone and Dolan (1977) and Rotblat (1981) provide extensive discussions of these. Here we draw upon those sources to estimate the dose burdens likely to be associated with the scenario.

A key factor is to estimate the unit-time reference dose for a particular location downwind from a detonation. This is defined as the dose rate (in rads/hr = rem/hr for these gamma ray emissions) at the time 1.0 hr after detonation. Table 23 shows the relationship between the dose rate at that time and subsequent times. Thus, if a location had a dose rate of, say, 10 rads/hr at time 1 hr, its rate at the end of three days would be 0.06 rads/ hr.

Table 23. Relative Dose Rates from Fallout.[a]

Time (hr)	Dose rate (rad/hr)
1	1000
2	400
4	150
6	100
12	50
24	23
36	15
48	10
72	6.2
100	4.0
200	1.7
500	0.50
1000	0.24

[a]Information summarized from Glasstone and Dolan (1977) and Rotblat (1981).

Dose rates, however, are not the primary controlling factor for health effects; rather, total accumulated dose is most important, i.e., the integration under a dose rate vs. time relationship. The complicated rates of decay of the diverse mixture of radionuclides in fallout lead to an empirical relationship of cumulative dose vs. time, relative to the unit-time reference dose rate, as illustrated in Figure 10. From this, it can be seen that the accumulated total dose integrated until $t = \infty$ is about 9.3 times the unit-time reference dose; i.e., for a unit-time reference dose of 100 rads/hr, the total dose to a person remaining at that location indefinitely would be 930 rads. The curve also shows that after just three days of exposure to local fallout, 86% of the infinite dose would have been accumulated.

This relationship is the first element needed for establishing those areas downwind of a detonation that include lethal levels of radioactivity from local fallout. However, one complication is that the relationship is based on the presence of fallout at the location immediately after the explosion (i.e., the time axis begins at the time of detonation, not at the time of first arrival of the fallout). Thus, the cumulative dose at a location must be calculated by estimating the unit-time reference dose (even if no

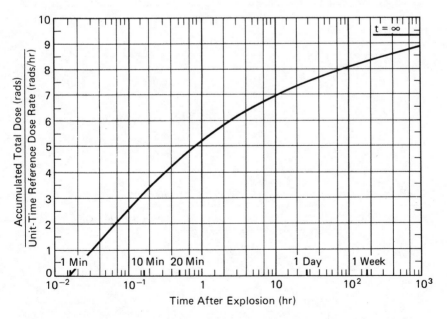

Figure 10. Accumulated total dose from local fallout as function of time after a nuclear detonation. (From Glasstone and Dolan, 1977)

actual radiation had arrived at t = 1 hr after the detonation), projecting the infinite dose, and reducing that value by the dose that would have accrued during the time period of zero until first arrival.

Calculation of the delay time prior to first arrival of fallout, however, cannot be done just by dividing the wind velocity by the distance to the specific location from the detonation. Rather, compensation must be made for the very fast radial expansion of the mushroom cloud itself, the source of the fallout. Wind is the effective transporter over the distance to a location from the nearest edge of the mushroom stem, not from the center of the cloud. For larger nuclear warheads, this can be an important phenomenon in speeding the arrival of early fallout (see Table 24).

Following Glasstone and Dolan (1977), we calculated the dimensions of the early fallout plumes for different warhead sizes using the following equations:

Table 24. Local Fallout Parameters[a].

A. *Parameters for Plume Calculations*

Unit-time reference dose (rads/hr)	c	d	y
3000	0.95	0.0076	0.86
1000	1.8	0.036	0.76
300	4.5	0.14	0.66
100	8.9	0.38	0.60
30	16	0.76	0.56
10	24	1.4	0.53
3	30	2.2	0.50
1	40	3.3	0.48

B. *Mushroom Cloud Radii for Land Surface or Low Air Bursts*

Warhead yield (kT)	Cloud radius (km)
100	2.4
200	3.6
300	4.8
500	6.1
1000	8.5
10,000	32.7

[a]Data from Glasstone and Dolan (1977).

$$L = 1.6 \ FcW^{0.45} \tag{26}$$
$$M = 1.6 \ FdW^y$$

where L = plume length (km)

M = plume width (km)

F = correction factor for wind velocity

$$= \left(1 + \frac{v - 24}{96}\right) \text{for } v \geq 24 \text{ km/hr}$$

$$= \left(1 + \frac{v - 24}{48}\right) \text{for } v \leq 24 \text{ km/hr}$$

v = wind speed (km/hr)

c = function of unit-time reference dose (listed in Table 24)

d = function of unit-time reference dose (Table 24)

W = warhead yield (kT)

y = function of dose yield (Table 24).

These calculations are presented in Table 25. A uniform wind speed of 24 km/hr (15 mph) was assumed, as in the Ambio Advisory Group (1982) scenario, with a wind shear of 15°. All calculations were made for surface bursts, with corrections made for mushroom cloud widths. Again, it should be made clear that the unit-time reference doses listed do not necessarily mean actual fallout dose rates at time 1 hr were as listed. Any distances greater than 24 km beyond the cloud radius (e.g., 8.5 + 24 = 32.5 km from ground zero for a 1 MT warhead) would have received no fallout at one hour.

To quantify the actual cumulative doses at a location, the entry time for the radiation was calculated using the distance from cloud margin to the location, divided by 24 km/hr. The long-term doses were then calculated by interpolating from Figure 11, and multiplying the resultant factor times the cumulated unit-time reference dose. Table 26 shows the results for cumulative doses near 450 rad, the LD_{50} for short-term exposure (taken here to be within first 48 hr after the detonation), 600 rad (the LD_{50} for long-term exposure), and about 1350 rad, based on a protection factor of 3 for reduction in external radiation by shielding (Rotblat, personal communication), resulting in a 450 rad dose at tissue for a person staying indoors during the immediate period. Thus, the first set of data apply to individuals unprotected during the early post-nuclear detonation period. However, it seems more reasonable to assume some actively sought protection via shielding from buildings, rubble, etc., so that the 1350 rad level corresponds to the appropriate lethal dose level.

Table 25. Unit-Time Reference Doses (rads/hr).

Plume length from ground zero (km)	100 kT[a]	200 kT	300 kT	500 kT	1 MT	10 MT
0	*	*	*	*	*	*
10	3.0×10^3	3.9×10^3	4.4×10^3	4.8×10^3	*	*
20	1.0×10^3	1.6×10^3	2.4×10^3	3.3×10^3	4.1×10^3	*
30	455	650	1.2×10^3	1.7×10^3	3.0×10^3	*
40	355	465	655	1.1×10^3	1.8×10^3	*
50	255	390	455	670	1.3×10^3	4.7×10^3
60	155	315	395	470	950	4.3×10^3
70	130	245	330	420	620	3.9×10^3
80	115	170	270	370	480	3.5×10^3
90	97	140	210	325	440	3.1×10^3
100	80	125	150	275	405	2.7×10^3
150	42	63	95	130	230	1.3×10^3
200	31	42	48	78	125	690
400	3.1	9.3	21	32	44	255
600	—	2.2	3.8	9.9	28	125
800	—	—	1.2	3.3	11.5	82
1000	—	—	—	1.2	4.1	48
1200	—	—	—	—	2.5	42
1400	—	—	—	—	1.0	36
4000	—	—	—	—	—	0.8

[a]Asterisks represent distance within mushroom cloud; dashes represent < 0.5 rads/hr; calculations from Eq. 26.

Considering the effects from other factors, such as blast, injury, psychological impact, age, and pre-war health, we consider this to be an actual lethal dose (LD_{100}). Similarly, for longer term health effects, 600 rad is considered to be the LD_{50} value (Rotblat, personal communication). Here we assume a protection factor of 1.0, since survivors would not remain for long periods of time in shelters and since over the long term, doses from inhalation of aeolian radioactivity and from ingestion of contaminated food and water would increase dose burdens beyond the calculations presented here. This, too, is taken to be an LD_{100} because of the complicating fators listed above. Thus, the population within the long-term 600 rad cumulative dose isopleth is considered to die from radiation exposure.

To calculate human health effects, we assume the plumes from each warhead detonation are independent and do not overlap. Obviously, the latter is not a realistic assumption for those situations, such as cities and ICBM sites, where many detonations could occur within a relatively small area. Thus, their plumes could have very large overlaps. However,

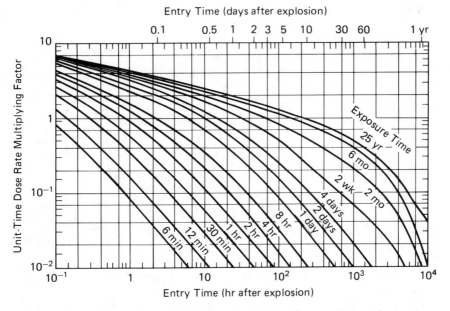

Figure 11. Accumulated total dose from local fallout from a nuclear detonation as a function of time of arrival of fallout at a location. (From Glasstone and Dolan, 1977)

unlike blast circles, fallout values are additive; for example, if at a location 100 rad cumulated dose results from one detonation and 50 rad from another detonation, a total of 150 rad exposure would ensue. Thus, to reach a particular lethal dose level, say, 600 rad for long-term dosage, two plumes of 300 rad each could overlap, or three of 200 rad, and so on. This additive phenomenon acts to increase the area covered by a lethal dose, whereas overlap of the 600 rad areas from each weapon acts to decrease total area. Therefore, assuming the area actually affected by 600 rad to be the sum of independent plumes, each contributing 600 rad, is a reasonable first approximation.

Values for local fallout plume coverage for the U.S. are given in Table 27. Assuming the protection factor of three, 12 million people could receive a lethal dose in the early post-war period, and 50 million over the longer term. [These values for U.S. fatalities from local fallout are comparable to the 67 million fatality estimate by Leaning (1983)]. Some of these fatalities coincide with blast and thermal radiation casualties from urban areas. However, the total area covered by long-term 600 rad doses is about one-fourth of the total U.S. land area, so using the value of

Table 26. Local Fallout Parameters.

Warhead yield (kT)	Plume[a] length (km)	Plume[a] width (km)	Arrival[b] time (hr)	Unit-time[c] reference dose (rad/hr)	Entry[d] factor	Cumulative[e] dose (rad)	Plume[f] area (km²)
450 rad/48 hr dose							
100	55	4.4	2.2	204	2.2	447	190
200	70	5.7	2.8	243	1.9	461	313
300	80	6.5	3.1	217	1.8	488	408
500	95	7.9	3.7	300	1.5	450	589
1000	130	12.4	5.1	299	1.5	449	1266
600 rad/long-term dose							
100	60	4.4	2.2	204	3.6	730	207
200	75	6.7	3.0	206	3.2	660	395
300	90	8.3	3.6	210	3.0	630	587
500	110	10.9	4.4	226	2.9	655	942
1000	150	16.7	5.9	228	2.6	595	1970
1350 rad/48 hr dose (450 rad dose × 3 protection factor)							
100	30	2.4	1.2	455	2.8	1275	56
200	40	3.8	1.5	465	2.7	1255	120
300	45	4.6	1.7	555	2.3	1275	163
500	50	5.8	1.8	670	2.2	1475	228
1000	65	8.6	2.3	785	2.0	1570	444

[a]From linear interpolation of Eq. 26.   [d]From Figure 11.
[b](Plume length − cloud width) ÷ 24 km/hr.   [e]Entry factor × unit-time reference dose.
[c]From Eq. 26.   [f]$A = \pi(\frac{1}{2}L)(\frac{1}{2}M)$.

average population density provides a reasonable estimate of additional fatalities from local fallout. Further, non-lethal radiation illness could affect additional millions, i.e., those with long-term doses as low as 200 rad.

These calculations represent an idealized case. Under realistic conditions, neither the plume nor the dose would be expected to be uniform. Geographic factors can significantly reduce dose. Hilly terrain, for example, may provide shielding of more than 50% (Glasstone and Dolan, 1977).

The meteorological conditions in the area of the burst and plume would affect both plume configuration and dosage. The area receiving an accumulated dose of ≥ 450 rads from a 500 kT burst varies by more than a factor of 3 between wind speeds of 16–72 km/hr. Precipitation is also influential. Residues become mixed and deposited unevenly (causing "hot spots") if rain originates above or within the mushroom cloud. Rain and snow clouds generally extend from a height of 3000 to

Table 27. Local Fallout Coverage.

A. *Air Bursts on Cities*

Target	Number of targets[a]	Warhead number–yield[b] (kT)	Total plume area[c] (km^2)	Population at risk[d]
		450 rad/48 hr dose		
Military bases[e]	239	1–300	97,500	
ICBM silos	1002	2–500	1,180,350	
Submarine bases	4	1–1000	5,060	
Total			1,282,910	32,073,300
		600 rad/long-term dose		
Military bases			140,300	
ICBM silos			1,887,880	
Submarine bases			7,880	
Total			2,035,980	50,898,525
		1350 rad/48 hr dose		
Military bases			38,950	
ICBM silos			456,900	
Submarine bases			1,775	
Total			497,625	12,441,125

B. *Surface Bursts on Cities*

Target	Total plume area[c] (km^2)	Population at risk[d]
	450 rad/48 hr dose	
Military (total)	1,282,900	
SMSA cities	384,600	
Ten cities	232,125	
Total	1,899,635	47,491,600
	600 rad/long-term dose	
Military (total)	2,035,980	
SMSA cities	535,300	
Ten cities	365,100	
Total	2,936,380	73,408,200
	1350 rad/48 hr dose	
Military (total)	497,625	
SMSA cities	148,240	
Ten cities	85,530	
Total	731,395	18,285,500

[a]From Center for Defense Information (1982) and Rand McNally (1983).
[b]From Table 3.
[c]From Table 26, assuming no overlap of plumes.
[d]Taken to be area × 25 people km^{-2} (average U.S. density).
[e]Includes army and air bases, naval bases, weapons storage facilities, DOE facilities, communication centers.

10,000 meters down to 1000 meters (Glasstone and Dolan, 1977). Thunderstorms, however, may have clouds up to 20,000 meters. Glasstone and Dolan (1977) conclude that for yields in excess of about 100 kilotons, precipitation scavenging should be insignificant unless thunderstorms are encountered. The height of the latter means that large portions of a 1 MT yield cloud could be overlapped. If a rain cloud encompasses the mushroom cloud, 50% of the radiation would be removed by rainout in .16 hours of rainfall; 99% would be removed in 1.1 hours. If the raincloud is above the mushroom, moderate rain would washout 50% of the radiation in 4 hr, 99% in 24 hr (Glasstone and Dolan, 1977). This would lead to substantial differences in the projected dose burdens at any location. Thus, the even distributions of radiation doses described by the ovoid plumes here are not highly accurate except under the most idealized conditions. Nevertheless, we believe this is the best estimate that can be made for anticipated radiation doses to humans.

One additional point about local fallout is the use of nuclear facilities as targets. Relatively small yield detonations occurring near nuclear power and weapons facilities (e.g., power plants, recycling plants, Pu reactors, storage sites) can result in exceedingly large local fallout. For example, a typical light-water reactor nuclear power plant could release the fallout equivalent to a 50–100 MT warhead, yet that could be initiated by a weapon of only tactical size (few kiloton yield). This phenomenon has been treated in depth in Ramberg (1982), and discussed in the Ambio Advisory Group (1982) analyses. The important factor for these targets is the long-term radiation left from the detonation, since nuclear power facilities have very large accumulations of long-lived radioisotopes from the fission process. Of particular importance are ^{137}C, ^{90}Sr, and ^{131}I in terms of their potential biological effect. The maps provided in Ambio Advisory Group (1982) indicate a very large coverage of the U.S. to the level of 100 rad by detonations over the 70 or so operational nuclear power plants, and a greater coverage over Europe. This dose burden would be additive to the doses resulting from local fallout from surface bursts in general, reducing the LD_{50} value needed from the latter to about 500 rad. This would be further reduced by addition of global fallout, as discussed next.

Global Fallout

The long-term fallout is that portion (approximately 20%) of residual radiation that takes longer than 24 hr for deposition. A major contribution to delayed fallout is from air bursts, which provide essentially no local fallout. There the radioactive particles and gases do not coalesce on particulates entrained from the ground, as in the case of a surface burst, but, rather, ascend with the mushroom cloud to the upper atmosphere. There they become more stabilized and have longer residence times in the air, allowing much of the radioactivity to decay prior to recontact with the surface. Thus, only longer-lived isotopes are of primary importance for global fallout, particularly ^{137}C, ^{90}Sr, and ^{131}I. These are biologically important because of their functioning as analogs to K (Cs) and Ca (Sr), and because of strong concentration in thyroid tissues for I. (See Howell et al., 1975, for a series of articles on Cesium cycling in natural systems.)

The rate of deposition is controlled by the degree to which the tropopause is penetrated and injection into the stratosphere occurs, a function of weapon yield and burst height, and by the general lateral and vertical circulation patterns of the atmosphere. The origin of fallout particles involves the condensation of vaporized material in the fireball, but that process alone would result in particles too small to descend in significant quantities; it requires larger particles of material entrained in the fireball to provide condensation seeds (Adams et al., 1960). According to Rotblat (1981), a three-compartmental model of the atmosphere reasonably predicts the global fallout experienced from weapons testing. Yet, as discussed in Turco et al. (1983a), there may be an underestimation of the intermediate-time fallout because the increased particulate loadings projected from nuclear war-induced fires can increase the rates of dry deposition and washout of the dispersed nuclear dust. Their projections for the base scenario of 5000 MT are for chronic whole-body doses from external exposure to reach 20 rad on average for the Northern Hemisphere for the first few months after the war. However, since during that period most fallout would be largely confined to the mid-latitudes (30°–60°), the dose could be about three times as large for survivors in those areas. With the addition of internal doses via consumption of contaminated food and water plus an occasional exposure to local fallout, Turco et al. (1983a) predicted about 100 rads for mid-latitude populations as a basal level. These would have to be added to the widespread coverage of local fallout for the majority of people, who could not live outside local fallout areas. Thus, as the case for additive doses from detonations over nuclear power plants, another 100 rad or so would be experienced by the survivors in the U.S. In short, the background dose burdens for survivors in the U.S. would be much higher than previous estimations, and widespread health effects, including fatalities, could be expected from residual radiation, including effects of chronic doses of radiation resulting in long-term cancer rate increases and perhaps even genetic effects. For a discussion of long-term radiation health effects, see Ishikawa and Swain (1981) and Glasstone and Dolan (1977).

Summary of Immediate Effects

The preceding analyses provide a basis for estimating the number of people in the United States who are killed or injured as a result of a large-scale nuclear war. These values are compiled in Table 28, reflecting a range of projected fatalities of 125–170 million people and an additional 30–50 million injuries. Thus, from the base scenario analyzed here, about 10–75 million Americans would survive the immediate effects of the nuclear war without direct, physical trauma.

These results should be tempered somewhat. They are based on a particular range of values for the scenario, in which about 110 million people are within targeted cities. Changing the urban targeting scheme would directly change that value and the associated casualty figures, so that targeting the smaller cities, for example, could increase the population at risk. Inspection of the data here suggests that adding smaller cities to targeting would increase the blast casualty figures disproportionately, since a higher percentage of the populations of smaller cities live within lethal areas. Conversely, targeting cities only above a large threshold of size would directly lower casualties from blast, thermal radiation, and fires; however, it would not significantly reduce casualties from local fallout, since most, if not all, surface bursts are associated with military targets. At the extreme of this, i.e., with no urban areas targeted, 40 million people could still die from the direct effects of local fallout.

To summarize some of the parametric analyses, changing urban targeting from all air to all surface bursts would have a net effect of reducing casualties by about 10 million. Daytime attacks vs. nighttime attacks would have the effect of increasing urban blast casualties by about 15%. This is based on U.S. census data (U.S. Bureau of the Census, 1983), which show 10% increases in urban populations for cities greater than 1×10^6, 15% for cities of 0.5–1.0×10^6, and 25% for cities under 500,000 in population. Daytime attacks could increase thermal radiation casualties by 20 million.

Local fallout is projected to cover about one-fourth of the U.S. continental area at long-term lethal levels (600 rad), not counting the inputs of perhaps 100 rad across most of the U.S. from global fallout,

Table 28. Summary of Immediate Effects.

Effect	Fatalities ($\times 10^6$)	Injuries ($\times 10^6$)	Total casualties ($\times 10^6$)
Blast[a]	50–80	30	80–110
Thermal radiation[b]	0.4–14	0.3–12	0.7–26
Fires[c]	0.7–7	0.7–7	1.3–13
Initial ionizing radiation[d]	0	0	0
Local fallout[e]			
Early doses	12–18	—	12–18
Local fallout[f]			
Long-term doses	39–55	—	39–55

	Air bursts on cities	Surface bursts on cities
Total Fatalities ($\times 10^6$)[g]	132–167	124–154
Injuries ($\times 10^6$)	30–50	

[a]From Table 9; includes those within 5 psi isopleth who die from other causes (e.g., fires), but all are treated as "blast" effect.
[b]From Table 11; does not include those already killed from blast.
[c]From Table 13; does not include those already killed from blast.
[d]All casualties from initial ionizing radiation would already be killed from blast.
[e]From Table 27; includes some people also covered in blast lethal areas, but also includes underestimate because of population distribution patterns; based on LD50 of 450 rad with protection factor of 3.
[f]From Table 27; includes population exposed to 600 rad long-term dose but minus those who die from early doses.
[g]Range of values for blast relate to all surface bursts (low value) to all air bursts (high value); range for local fallout is all surface bursts for cities (high value) to all urban air bursts (low value). Thus total range includes high blast plus low fallout figure and vice versa. Ranges for other effects are independent; for thermal radiation, range reflects day (high)/night (low) differences; for fires, range reflects variance in fuel densities. One other day/night effect is included here: urban effects can affect 15% more people during the day as urban populations increase from census (residential) values.

nuclear power plant disruptions, and internal dose routes. Over the long term, very few survivors would not accumulate doses in the few hundred rem as they travel, consume food, and otherwise become exposed to residual radioactive debris. Further, major portions of the U.S. land would remain contaminated for long periods of time.

Clearly, the immediate effects of a large-scale nuclear war could be devastating for the United States. Studies by the World Health Organization (Bergstrom et al., 1983) for a similar scale nuclear war suggest 1.1 billion deaths and 1.1 billion additional injuries worldwide. This is supported by the estimate of 750 million deaths worldwide according to the *Ambio* analyses (Middleton, 1982). These values set the stage for the considerations of the intermediate and long-term effects on humans and natural systems.

4

Intermediate and Long-Term Consequences

From the preceding analyses, it is clear that a large number of human casualties would occur in the immediate period after multiple nuclear explosions; that the radioactive fallout would inevitably deposit at the surface at high levels of activity across large scales of landscape; and that the inputs into the atmosphere of particulates and other emissions from secondary fires would result over the subsequent weeks and months in extraordinary changes in weather and insolation. From this basis, we can now look forward in time to the repercussions of these phenomena on human life.

This chapter highlights a central question of this book: Given the scenarios being analyzed, given the resultant atmospheric and climatic impacts, given the direct effects on humans and the environment from a large-scale nuclear war, what does this mean for humans and ecosystems in the weeks to decades after the nuclear war? In short, what are the human health/societal/ecological responses to this catastrophic global insult. Obviously, much of what could be said about such a post-war Earth is highly speculative, and uncertainties abound. Nevertheless, the nature and magnitude of the environmental changes predicted by Turco et al. (1983a, b) are such that many aspects of ecological damage resulting from extreme conditions are rather obvious, even if difficult to quantify with certainty (Table 29). By systematically examining these stresses and the mechanisms for responses by human and ecological systems, a picture emerges sketching the reality of a new world that humans potentially could enter in a flash of time.

Based in particular on the new results from the Turco et al. (1983a,b) study, it is clear that some of the environmental conditions may be so extreme as to cause totally catastrophic effects; i.e., there is a set of phenomena, each of which alone would be adequate to cause human and ecological destruction on a global scale. This chapter will first address those aspects as separable elements in the response patterns.

Table 29. Long-Term Stresses[a] on the Northern Hemisphere (N.H.) Biosphere

A. *Following a 5000 MT Base-Case Nuclear War*

Stress	Perturbed value[b]	Duration after nuclear war	Area affected	Range of uncertainty
Sunlight intensity	× 0.03	1 wk	N. Midlat.	× 0.01–0.1
	× 0.1	1 mo	N. Midlat.	× 0.03–0.3
	× 0.5	2 mo	N.H.	× 0.1–0.9
	× 0.8	4 mo	N.H.	× 0.3–1.0
Land surface temperature[c]	−18°C	1.5 mo	N. Midlat. land	−33° to −8°C
	−3°C	3 mo	N. Midlat. land	−18° to −3°C
	+7°C	10 mo	N.H. land	+2° to +13°C
UV-B radiation[d]	× 1.5	1 yr	N.H.	× 1.2–2.0
	× 1.2	3 yr	Global	× 1.0–1.5
Surface ozone[d,h]	150 ppbv	3 mo	N.H.	100–250 ppbv
Radioactive fallout exposure[e]	100–1000 rad	1 hr–1 day	3% N. Midlat. land	
	10–100 rad	1 day–1 mo	10% N. Midlat. land	Factor of 3
	< 10 rad	>1 mo	30% N. Midlat. land	
Fallout burdens[d,f]	[131]I 2 × 10⁵ MCi	8 day[g]	20% N.H.	
	[106]Ru 5000 MCi	1 yr	N.H.	Factor of 2
	[90]Sr 200 MCi	30 yr	Global	
	[137]Cs 330 MCi	30 yr	Global	

B. *Following a 10,000 MT Base-Case Nuclear War*

Stress	Perturbed value[b]	Duration after nuclear war	Area affected	Range of uncertainty
Sunlight intensity	× 0.01	1 wk	N. Midlat.	× 0.001–0.1
	× 0.05	3 wks	N. Midlat.	× 0.01–0.5
	× 0.25	1.5 mo	N.H.	× 0.1–1.0
	× 0.50	3 mo	N.H.	× 0.3–1.0
Land surface temperature[c]	−18°C	2 mo	N.H. land	−33° to −8°C
	−3°C	6 mo	N.H. land	−23° to −3°C
	+7°C	1 yr	N.H. land	−13° to +13°C
UV-B radiation[d]	× 5	1 yr	N.H.	× 3–10
	× 3	3 yr	N.H.	× 2–5
	× 2	3 yr	N.H.	× 1–3
Surface ozone[d,h]	150 ppbv	3 mo	N.H.	100–250 ppbv
Radioactive fallout exposure[e]	100–1000 rad	1 hr–1 day	5% N.H.	
	10–100 rad	1 day–1 mo	20% N.H.	Factor of 3
	<10 rad	>1 mo	50% N.H.	
Fallout burdens[d,f]	[131]I 4 × 10⁵ MCi	8 day[g]	20% N.H.	
	[106]Ru 1 × 10⁴ MCi	1 yr	N.H.	Factor of 3
	[90]Sr 400 MCi	30 yr	Global	
	[137]Cs 650 MCi	30 yr	Global	

Table 29. *(continued)*

C. *Following a 10,000 MT Worst-Case Nuclear War*

Stress	Perturbed value[b]	Duration after nuclear war	Area affected	Range of uncertainty
Sunlight	× 0.01	1.5 mo	N. Midlat.	× 0.003–0.03
intensity	× 0.05	3 mo	N. Midlat.	× 0.01–0.15
	× 0.25	5 mo	N.H.	× 0.1–0.7
	× 0.50	8 mo	N.H.	× 0.3–1.0
Land surface	−43°C	4 mo	N. Midlat. land	−53° to −23°C
temperature[c]	−23°C	9 mo	N.H. land	−33° to −3°C
	−3°C	1 yr	N.H. land	−13° to +7°C
UV-B	× 4	1 yr	N.H.	× 2–8
radiation[d]	× 3	3 yr	N.H.	× 1–5
Surface ozone[d,h]	150 ppbv	3 mo	N.H.	100–250 ppbv
Radioactive	100–1000 rad	1 hr–1 day	5% N.H.	
fallout	10–100 rad	1 day–1 mo	20% N.H.	Factor of 3
exposure[e]	<10 rad	>1 mo	50% N.H.	
Fallout	^{131}I 4 × 10^{5}MCi	8 day[g]	20% N.H.	
burdens[d,f]	^{106}Ru 1 × 10^4 MCi	1 yr	N.H.	Factor of 3
	^{90}Sr 400 MCi	30 yr	Global	
	^{137}Cs 650 MCi	30 yr	Global	

[a]Data compiled from Turco et al. (1983b) and Turco (personal communication). Stresses occur simultaneously. Their geographic extent and severity would depend on many factors, including: number, distribution, and yields of weapons detonated; height of bursts; scale of subsequent fire spread; degree of atmospheric transport of soot and dust (especially for transport from N.H. to S.H.); rate of washout of soot and dust, as affecting atmospheric residence times.

[b]The following abbreviations apply: 'x' indicates a multiplicative factor; ppbv = parts per billion by volume; MCi = megacurie.

[c]Average surface temperatures should be compared to the ambient value of 13°C for the Northern Hemisphere.

[d]From Turco et al. (1983a,b), Crutzen and Birks (1982).

[e]These figures are rough estimates of whole-body gamma ray doses and apply only to exposed organisms. Doses are from "prompt" and "intermediate" fallout; ingestion of biologically active radionuclides is not taken into account. They could double the dose in body organs (e.g., the thyroid for ^{131}I), where these radionuclides tend to accumulate.

[f]The principal modes of deposition are fallout and washout. In air bursts, the radionuclides settle out slowly over several years. In surface bursts, ~60% falls out promptly, ~40% over 1–2 years (Glasstone and Dolan, 1977). In subsurface water bursts, ~100% is deposited in the water. During the atmospheric nuclear tests of the 1950's and 1960's, ~200 MT of fission yield produced an average ^{90}Sr deposition ~50 millicuries/km^2.

[g]These are essentially the radionuclide half-lives. Other radionuclides contribute mainly to the prompt fallout exposure.

[h]Ozone generation in smog reactions depends sensitively on the gases generated by the fires. The numbers quoted here are estimates based on Crutzen and Birks (1982).

The latter half of this chapter will discuss other problem areas, each of which would be considered unacceptable if occurring without all the other effects of nuclear war, but which is not so serious as to call into question the survivability of humans and other species.

Major Problem Areas

The consequences from the initial conditions study that are likely to be totally catastrophic are climatic changes relating to temperature, atmospheric changes regarding incident sunlight, radiation levels from global and local fallout (discussed previously), impacts on the food-producing capabilities of the globe, and collapse of societal structures and functions. Each will be treated separately here.

Effects of Reduced Temperatures

Human Response to Cold. The lethal cold limit for humans depends on variables like metabolic activity, size, physical fitness, and accustomed temperature as well as moisture level and wind speed. As in all species, heat loss is proportional to the amount of surface area per unit of body mass. Humans are basically tropical species, however, and rapidly require protection for survival in low temperatures.

Most investigations of temperature limits have focused on the effect of short-term exposures of cold on portions of the body (e.g., hands, face, fingers). For whole-body exposures, however, the unit *clo* has been developed as a measurement of insulation. One clo is defined as the insulation provided by an indoor business suit. It will maintain a skin temperature of 31°C when the air is at 21°C and wind is minimal (Bell et al., 1980). A naked adult might have an average insulation of 1.6 clo from tissue and still-air insulation. Under these conditions, his/her metabolism at rest would have to double for each drop of 8°C to maintain comfort (human metabolisms can increase only 3 times for sustained periods). Conversely, a heavily clothed human has about 6 clo and would have to double his/her metabolism for drops of 26°C (Burton and Edholm, 1955).

Burton and Edholm (1955) conclude that the lowest environmental temperature that could be sustained indefinitely by a naked person is about 2°C. Edholm (1978) estimated that air temperatures of 5 to −2°C combined with damp conditions would rapidly cause hypothermia.

Conversely, LeBlanc (1975) suggested lower limits of −15 to −20°C, and humans have reportedly survived air temperatures of −28.9°C when walking at 3.5 mph (Altman and Dittmer, 1966; this report from a 1949 study by H.S. Belding contains no information as to duration of experiment or clothing worn). The differences between these estimates may be the length of exposure.

Non-lethal cold has negative effects, especially on people who have lost blood and/or are not mobile. Frostbite of appendages (which may become gangrenous) can occur at 0°C; damp conditions enhance this effect (Killian, 1981). Poor nutritional conditions will also increase susceptibility. Additionally, cold is likely to reduce the rate of healing because the skin blood flow is decreased (Edholm, 1974). Sound sleep is impaired at internal temperatures of about 35°C (Killian, 1981). The warming of severely chilled people may not save them, and the stress involved may cause cardiac arrest.

Humans, like other animals, can metabolically adapt to prolonged exposure to moderate cold (rats adapt within a few weeks). Examples of this are seen in Australian aborigines, Kalahari bushmen, and Korean women divers (LeBlanc, 1978). Neither the time required for adaptation nor the temperature limits possible are known, however.

It seems likely that only well-sheltered or otherwise protected people could withstand prolonged exposure to −25°C. Appropriate clothing, kept dry, would probably permit life at 0 to 10°C. Clearly, under the conditions of temperature reductions projected by Turco et al. (1983a,b), initially surviving humans would experience lethal temperatures that could only be mitigated by considerable sheltering, energy usage, insulating clothing, and related conditions. The potential is for wide-spread fatalities from exposure to cold, particularly for those already injured and for elderly and young individuals.

Physiological Responses of Biota. The consequences of dramatically reduced temperatures on the ecosystems of the Earth would stem from direct effects of low temperature on the animals and plants of a given region, and from secondary effects on their food and water supplies. The effects of large reductions in temperatures on plants depend on the time of year, the duration of low temperatures, and the tolerances of plants to reduced temperatures.

Plants differ dramatically in their tolerance limits, but the plants of a given region generally have rather similar limits (Table 30) that are related to the normal extremes of the low temperatures of their native environments. In the base case scenario, the temperatures are expected to fall rapidly to levels that exceed the historic recorded lows in most

Table 30. Temperature resistance of the leaves of vascular plants from different climate regions.[a]

Plants	°C for cold injury in the hardened state
Tropics	
Trees	+5 to −2
Forest undergrowth	+5 to −2
Mountain plants	−5 to −10
Subtropics	
Sclerophyllous woody plants	−8 to −12
Subtropical palms	−5 to −14
Succulents	−5 to −10
C_4 grasses	−1 to −3 (−8)[b]
Temperate zone	
Evergreen woody plants of coastal regions with mild winters	−6 to −15 (−25)[b]
Arcto-tertiary relict trees	−10 to −25 (−15 to −30)[b]
Dwarf shrubs of Atlantic heaths	−20 to −30
Winter-deciduous trees and widely distributed shrubs	(−25 to −40)[b]
Herbs	
Sunny habitats	−10 to −20 (−30)[b]
Shady habitats	
Water plants	ca. −10
Cold-winter areas	
Evergreen conifers	−40 to −90
Boreal broad-leaved trees	(−196)[b]
Arctic and alpine dwarf shrubs	−30 to −70
Herbs of the high mountains and arctic	(−30 to −196)[b]

[a]Limiting temperatures are for 50% injury (TL_{50}) after exposure to cold for 2 hr or more, or after exposure to heat for 0.5 hr. The data were taken from many original publications. Information compiled from Altman and Dittmer (1973), Larcher (1980), Larcher and Bauer (1981).

[b]Vegetative buds.

regions within about two weeks (Figure 8). The abrupt onset of cold is of particular importance because cold-tolerant plants normally require some pre-conditioning to be fully tolerant of cold. Winter wheat, for example, can tolerate temperatures as low as −15° to −20°C when pre-conditioned to cold temperatures (as occurs naturally in fall and winter months), but may be killed by temperatures of −5° to −10°C occurring in summer (Larcher and Bauer, 1981). Dormant buds of deciduous trees of the boreal forest zone can even tolerate cooling to liquid nitrogen temperatures. It is, therefore, unlikely that these plants could be killed by any conceivable low temperatures reached in natural environments if these occur during winter, but the same species can be quite sensitive to freezing at −10°C if this occurs while the plants are actively growing.

Nearly all terrestrial plants in the continental regions of the Northern Hemisphere would be killed or severly damaged by cold if the nuclear exchange were to occur just prior to or during the growing season. Plants of the northern latitudes would most likely be cold hardened and have a higher probability of surviving the cold temperatures if the nuclear war were to occur in the fall or winter. Approximately one third of the Earth's land area never experiences freezing temperatures, and a significant additional area does not experience temperatures below −10°C. Plants in these areas generally do not possess dormancy mechanisms that would enable them to tolerate very low temperatures, and they would be especially sensitive to damage regardless of the timing of the war.

Even temperatures considerably above freezing can be damaging to some plants. For example, exposure of rice or sorghum to temperature of only 13°C at the critical time can inhibit grain formation because the pollen produced is sterile (Larcher and Bauer, 1981). Corn (Zea mays) and soybeans (Glycine max), two important crops of North America, are quite sensitive to temperatures below about 10°C.

The conclusion seems inescapable that temperatures predicted in the base case scenario would cause massive damage to terrestrial plants. Net primary productivity in most ecosystems would be temporarily halted until plants could resprout or be replaced from seed when favorable conditions returned. The only area in which plants might not be devastated by severe cold would be immediately along the coasts and on islands where the temperatures would be moderated by the thermal inertia of the oceans. Other stresses to plants from radiation, air pollutants, and low light levels (Table 29) that follow the war would compound the damage from freezing. In addition, diseased or damaged plants have a reduced capacity to acclimate to low temperatures (Larcher and Bauer, 1981; Levitt, 1980).

There are numerous examples suggesting that unusually low temperatures from nuclear war could cause widespread mortality of animals. Precht (1973) cites examples of abnormally cold winters bringing about total annihilation of some species. Some invertebrate animals survive freezing (examples include arctic insects and mollusks of the intertidal region). Vertebrates either do not survive freezing (excepting special cryogenic preservation procedures), or at least die later of the consequences (Precht, 1973). Some animals have adaptations to permit supercooling to temperatures below the freezing point without actual ice formation. But many other animals die at body temperatures that are well above freezing (Table 31).

Table 31. Lower Temperature Limits for Some Thermoregulating Animals.[a]

Species	Lowest body temperature[b] (°C)	Lowest ambient temperature tolerated for 1 hr without hypothermia (°C)
Human	24–26	−1 (naked)
Guinea pig	17.5–21	−15
Fowl	23	−50
Rat	13	−25
White fox	—	−80
Duck	—	−100

[a]Compiled from Precht (1973).
[b]Without medical revival.

Injury to the respiratory center or breakdown of osmoregulation is generally cited as the cause of death at chilling temperatures. The well-known sensitivity of tropical fish to temperatures of 10° to 15°C is an example of this form of response. The thermal sensitivity of a particular animal may vary considerably during its life, generally being more tolerant in normally cold seasons. Reproductive functions are also quite sensitive to cold, and inhibition of this by prolonged cold could in itself cause annihilation of short-lived animals.

Homeothermic animals can maintain their body temperature under adverse temperatures through the metabolic generation of heat. There are limits to this process related to the thermal insulation of the animal and its metabolic reserves. Ducks can survive −40°C for 7–16 days without freezing, while doves can live for 2–6 days under similar conditions. Their glycogen reserves are exhausted in the first 8 hours, but body temperature is maintained until death occurs, primarily as a result of starvation rather than freezing (Hensel et al., 1973). In this regard the impact of low temperatures may be exacerbated by darkness, which would inhibit feeding. The lowest ambient temperature tolerated without hypothermia (Table 31) provides some relative index of the abilities of various animals to thermoregulate.

In general, translating the *physiological responses* of individual plant and animal species into *ecosystem-level responses* is not easily or directly accomplished. Stress responses of individuals may differ from population-level responses because of the genetic and phenotypic differences among individuals and because of the potential for compensatory mechanisms (Levin et al., 1983), particularly if density-dependent phenomena are involved. Further, the issue of spatial and temporal scales is critical to the population-and community-level responses; for example, extreme stresses that cover only a fraction of the areal distribution of a species may result in local extirpations, but recovery can follow from reintroduction of the species into the affected areas from outside refugia. Many other examples could be given of the importance of the scale of a stress relative to the characteristic scales of the stress recipient.

In the present situation, however, density-independent forces would predominate over scales of continental size or greater. That is, the magnitude and extent of the adverse physical conditions would be so large that essentially all individuals of a biological population would be subject to the stresses. Thus, the opportunity for population-level compensation from adverse biotic responses would be very limited, and the estimates of physiological effects on individuals can reasonably confidently be extrapolated directly to higher levels of biological organization, i.e., to ecosystems. That is not to say that strategies could not be found by certain species for assuring their survival nor that certain classes of structural or functional groups of species would not have a relatively increased capacity for resilience in response to extreme stress, in this case, sudden onset of extreme cold. [For example, plants with buried seed sources (geophytes) might have lesser long-term impacts compared to, say, plants living completely aboveground (epiphytes).] But this does mean that some of our understanding of stress ecology may not apply to the case of hemispheric-scale perturbations.

Beyond the scale question, evaluating ecosystem responses is made difficult by the lack of empirical data. There simply have not been many experiments in which whole ecosystems or even representative microcosms have been subjected to sudden decreases of temperatures from an ambient level to $-20°C$ or so. Such a phenomenon has not heretofore been considered a plausible anthropogenic stress on natural systems, so there has been little impetus for this type of study. Perhaps that should be an area of study in the future. For the present the best source of understanding ecosystem responses is by simulations of ecosystem models that include the effects of temperature and light levels on system productivity and structure. It must be kept in mind, however, that such simulations carry the models outside the bounds for which they were designed to be valid, and, therefore, substantial uncertainty exists in specific outcomes. Yet models can help elucidate the interactions among components of systems too complex for ready understanding.

Grassland Ecosystems. The consequence analyses on ecosystem effects involved the usage of two computer models, simulating short-grass prairie and eastern deciduous forest ecosystems. The grassland model

(SPUR) was utilized by staff at the U.S.D.A. Agricultural Research Service and the Natural Resource Ecology Laboratory, Colorado State University, both in Fort Collins, Colorado. SPUR is a comprehensive rangeland simulation model with climatic, hydrologic, plant, animal, insect, and economic components (Wight, 1983; Wight et al., 1983), designed for simulations on scales from local pasture and grazing units to the basin level. Of primary concern here is the plant growth component (Hanson et al., 1983), which uses a difference-differential equation model simulating the production processes controlling plant biomass and nitrogen flows. Abiotic driving variables include air temperature, precipitation, soil water potential, solar radiation, and total daily wind. The model was calibrated to representative environmental conditions in a grassland ecosystem in Colorado, including inputs describing current ambient temperature, sunlight, and precipitation regimes. The model included the components characteristic of such an ecosystem. Specifically, daily productivity of five compartments was simulated, representing (1) warm season grasses (C4 photosynthesis); (2) cool season grasses (C3 photosynthesis); (3) warm season forb species (not necessarily C4); (4) cool season forbs (typically C3); and (5) dwarf shrubs (C3). Weather data for the control simulations consisted of actual data for 1974 and 1975 (for simulation years 1 and 2, respectively).

The major output variables monitored were: (1) net aboveground primary production (in units of g dry wgt m^{-2} yr^{-1}); (2) net belowground production (g m^{-2} yr^{-1}); and (3) peak aboveground standing living biomass (g m^{-2}).

The environmental conditions for the control simulation followed actual data using a 1-day time step. Temperature decreases were effected by decreasing the daily values sufficiently that the average annual temperature reductions were 3°, 6°, and 9°C below normal ambient. This was designed to reflect the longer term response of the ecosystem to changes in climate beyond one year after the nuclear war. Conditions within the initial year after a nuclear war would be so extreme that no aboveground or belowground productivity would ensue, and all standing green biomass at the beginning of the war would soon be dead. Thus, the first year's simulations began with assuming the prairie ecosystem was just emerging from a typical mid-winter condition, and reflect conditions 1–2 years *after a nuclear winter*.

It is important to note that the probable decrease in precipitation to continental regions projected by Turco et al. (1983a) was not included in these analyses. This could have a significant effect on the simulations, since grasslands in general are moisture-limited ecosystems. As seen by the effects of relatively small decreases in annual precipitation on grassland and crop ecosystems, this effect could be very important. Also, these simulations did not include the effects of radiation (not expected to have widespread consequences on grasslands), aeolian inputs of dust, increased soil erosion, fires, and so on. The effects of longer term reductions in insolation were simulated, as discussed in a later section.

Results from the temperature reduction simulations are summarized in Table 32. Temperature reductions of 3°C cause less than a 10% decrease in productivity and standing biomass, 6°C leads to a 25% reduction, and 9°C, 40–50% decreases. Assuming the model simulations are reasonably accurate, this suggests that natural grasslands, the closest analog to crop systems, are less sensitive than the latter to external

Table 32. Summary of SPUR Grassland Model Simulations.

A. *Temperature Reductions*

		Control	Temperature reductions		
			3°C	6°C	9°C
Net primary	Year 1	100	91	72	58
production	Year 2	100	87	67	49
(% of control)					
Peak					
aboveground	Year 1	100	92	76	65
standing					
live	Year 2	100	93	76	60
biomass					
(% of control)					

B. *Light Reductions*

		Control	Insolation reductions		
			25%	50%	75%
Net primary	Year 1	100	89	76	61
production	Year 2	100	96	89	82
(% of control)					
Peak					
aboveground	Year 1	100	90	78	64
standing					
live	Year 2	100	94	87	75
biomass					
(% of control)					

perturbations. On the other hand, crop productivity is important in terms of yield of grain, not the amount of living biomass standing aboveground. Generalizations here are tenuous.

Forested Ecosystems. To evaluate the effects of temperature and light reductions on forested ecosystems, the FORNUT model (Weinstein, 1983) was utilized. This model is implemented at Cornell University and is being adapted for simulations of effects of air pollution on forested ecosystems (Harwell and Weinstein, 1982; Weinstein et al., 1983; Weinstein and Harwell, 1984). The FORNUT model version used for the present study is calibrated to the mixed conifer-hardwood forest of the Southern Appalachian Mountains, based on stand data from an area typical of eastern Tennessee. Simulations indicate that in the absence of strong perturbations, the forest structure in this region will approach a steady-state in approximately 50 years.

The model is based on the forest gap replacement concept in which a single simulated plot covers one-twelfth hectare, about the size of the canopy of a very large tree. Simulations monitor the birth, growth, and death processes for individual trees selected randomly from an available seed source. These processes have both deterministic components, as, for example, optimal growth curves characteristic of each species, and stochastic aspects, including random occurrence of tree sprouting, blowdown, and death. By simulating a large number of individual plots, each with its own stochastic processes subject to the same external driving parameters as the other plots, a picture emerges of the species composition, age structure, size classes, and successional patterns of the forested ecosystem as a whole.

Tree production is simulated as a function of available sunlight and temperature, which affect both the decrement from optimal growth conditions for each species (based on degree-days over the growing season) and the rates of evapotranspiration experienced by the plot, with concomitant influence over the soil water availability. In actual ecosystems, the structure and dynamics of a forest are affected by interspecific competition between trees for light, water, and nutrients. FORNUT is designed explicitly to include this inter-specific competition factor; thus, its use for evaluating the effects of reduced light and temperatures on forested ecosystems provides both a look at the direct effects on particular species and a look at structural changes resulting from differential sensitivities of tree species and associated changes in competitive interactions.

A series of simulations was conducted using the validated version of the model and actual weather data from Oak Ridge, Tennessee, as the control. Temperature and light reductions were done as with the SPUR model. However, unlike the grassland situation, where the first year after the nuclear war could be assumed to have no productivity and no living standing biomass and, thus, initial conditions reflected emergence from a typical winter condition, the forested ecosystem maintains most of its living structure as biomass carried over from year to year. The initial year after the nuclear war, though, could experience climatic conditions that could kill most of the perennating tissues of the trees depending on the time of the year of the war, thereby creating a forest of standing dead that would enter a subsequent period of more benign conditions. There is no way of estimating the state of the ecosystem at that point through the model simulations, since such conditions are far outside the bounds of applicability. We have ignored that problem and initialized the model as if the forest were suddenly in a five-year period of somewhat decreased light and temperatures, but not having just experienced months of extreme conditions. Additional parametric simulations indicate that this assumption does not greatly affect the predictions.

Results from the FORNUT simulations indicate that temperature reductions during the 2–6 year period after a nuclear war, with subsequent return to pre-war levels, would create significant changes in forest ecosystems over long periods of time. These changes are manifested in terms of both the total living biomass in the forest and the community structure of the tree species.

Figure 12 represents the effects on total biomass. Four comparisons are made, with temperature reductions under pre-war light and precipitation regimes of 0°, 3°, 6°, and 9°C below the average annual levels. The 0°C reduction constitutes the control simulation, demonstrating the successional development of the forest over a fifty-year period, beginning with the forest as currently constituted. During that period, the total biomass of standing, living trees increases from about 80 metric tons ha^{-1} to a value of about 130. Fluctuations in the simulation result from stochastic variation in the birth, death, and growth rates of individual trees; twenty plots were simulated for this and each other case to reduce that variability and to result in a reasonable picture of the dynamics of the entire forest, rather than of just a single plot.

The simulations of a 3°C reduction in temperature over a five-year period show an initial decrease in biomass by about 25%, but within three or four decades the biomass has returned to the unstressed trajectory. By contrast, a reduction in average air temperatures of 6°C for five years results in initial biomass losses of 80%, peaking a few years after the temperatures return to pre-war levels. Further, the simulations show

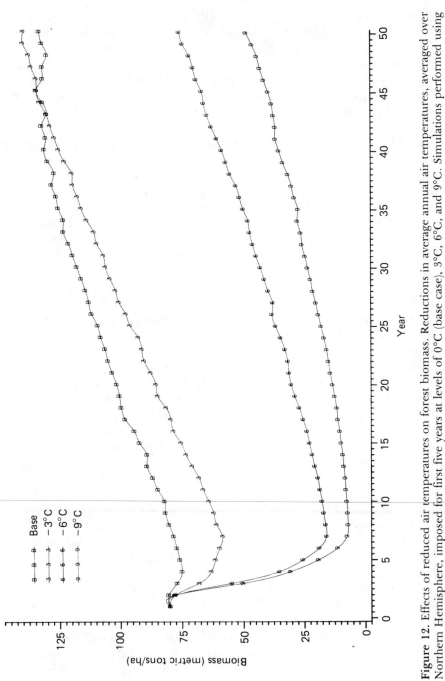

Figure 12. Effects of reduced air temperatures on forest biomass. Reductions in average annual air temperatures, averaged over Northern Hemisphere, imposed for first five years at levels of 0°C (base case), 3°C, 6°C, and 9°C. Simulations performed using FORNUT model.

that after the nuclear war, if a period of 6°C reductions occurs, even fifty years later the total forest biomass would be below initial values and only half of what the forest would have had resulting from normal forest development. A similar pattern results from a five-year decrease by 9°C, with peak biomass losses of 90%. Other simulations using different initial state of the simulated forest (not shown) confirm the patterns seen here, indicating little impact on the results from starting with different initial forest structures.

Another set of simulations were performed to test the effect of the co-occurrence of precipitation reductions with temperature reductions (Figure 13), in this case a 6°C decrease. As before, the first year's extremes of climate were ignored, and only the effects of five years of temperature and precipitation reductions were examined. Three simulations were run, with 0%, 10%, and 25% reduction in precipitation. Results suggest that the effect of 6°C reduction is considerably more consequential than a co-occurring reduction in water inputs to the forest.

A closer look at the forest changes resulting from temperature reductions is presented in Figures 14a and b. In these can be seen the dominant plant species at times 0 and 50 years, and comparing 0°, 3°, 6°, and 9°C reductions. The initial graph shows equal biomass within species across temperature reductions, reflecting common initial conditions for all simulations. However, by 50 years after the war (and 45 years since the climate returned to normal in the simulations), quite substantial differences in species composition exist. Comparing the control and 3°C simulations, only the species *Oxydendron arboreum* (sourwood) has dropped out of the forest by year 20 and remains gone thereafter; otherwise, little change in community structure among the dominants is evident. However, the 6°C reduction shows a relative increase in *Acer rubrum* (red maple), *Fraxinus americana* (ash), and *Prunus serotina* (black cherry), but a loss of *Carya glabra* (hickory), *Liriodendron tulipifera* (tulip poplar), *Nyssa sylvatica* (blackgum), *Oxydendron arboreum* (sourwood), and *Quercus prinus* (oak) biomass. The same pattern ensued from a simulated 9°C reduction.

The FORNUT simulations indicate that a few years of climatic change could have long-lasting impacts on the living components of an eastern deciduous forest of the lower Appalachians. Forests in more northern locations might fare better, having a seed source with greater representation of cold-adapted tree species; however, regional-scale patterns of climatic changes could subject more northern areas to even greater temperature reductions. Thus, it is difficult to extrapolate across mixed-deciduous forests at different locations and to lower diversity forests,

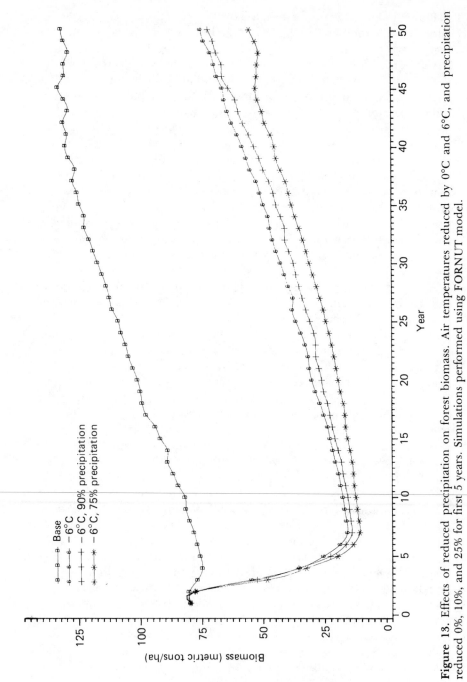

Figure 13. Effects of reduced precipitation on forest biomass. Air temperatures reduced by 0°C and 6°C, and precipitation reduced 0%, 10%, and 25% for first 5 years. Simulations performed using FORNUT model.

such as the coniferous forests of the Western U.S. The simulations are informative, nevertheless, in showing long- term changes in both forest size and structure.

Freshwater Ecosystems. Effects of reduced temperatures on freshwater ecosystems would be controlled by the degree to which surface water systems are subject to freezing. Using the atmospheric changes (Figure 8) projected from the 5000 MT exchange scenario of Turco et al. (1983a), we conducted analyses of the importance of freezing.

Two major atmospheric changes, the fall in temperature and the reduction in precipitation, would reduce the availability of fresh water to humans and other species. The most immediate consequence of the fall of temperature would be the extensive freezing of surface water bodies. Both the reduced amount of total precipitation and the fall of most or all new precipitation as snow or ice would reduce the amount of liquid water flowing into the frozen lakes and rivers, further reducing the availability of water to organisms. Figure 15 shows the general chain of effects.

The extent of flow reduction and freezing depends on the patterns of the hemispheric water cycle. The sections that follow discuss the distribution of fresh water in Asia, Europe, and North America and the possible extent of fresh water freezing. It is assumed that (1) the 5000 MT or the 10,000 MT scenario occurs as described in Turco et al. (1983a, b) and as cases 1 and 9 in Figure 8; (2) normal seasonal temperature and water storage variations are minor relative to the effects of the detonations.

Lakes store most of the fresh water on the surface of the Earth. Rivers and streams hold much less water at once. Table 33 inventories average river and lake storages and river flows for the three Northern Hemisphere continents.

River storages change rapidly with varying flow. Data in Table 33 on the source of river flow suggest that an elimination of overland runoff could reduce river flows by two-thirds to three-quarters. Reductions in total and liquid precipitation would translate into quick (probably within a month) and significant reductions in river flows and storages.

Lake storages are more static. Their long turnover times compared to rivers and streams make them less sensitive to short-term precipitation losses. Smaller lakes and ponds are very numerous on all of the continents, but most of the total lake water volume is tied up in a few large lakes (Table 34). Lake Baikal in Asia and the Great Lakes of North America together hold about 80 percent of the 55,000 cubic kilometers of fresh lake water in the Northern Hemisphere.

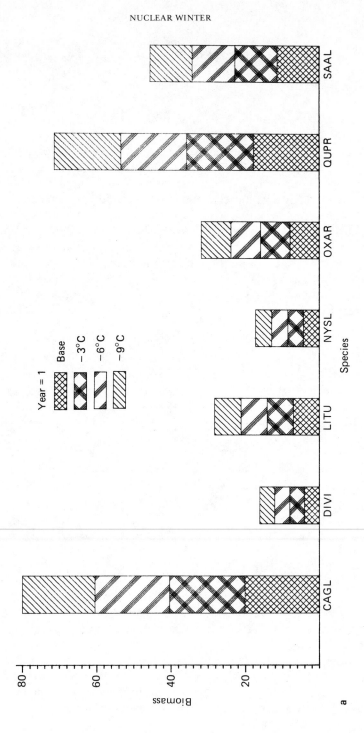

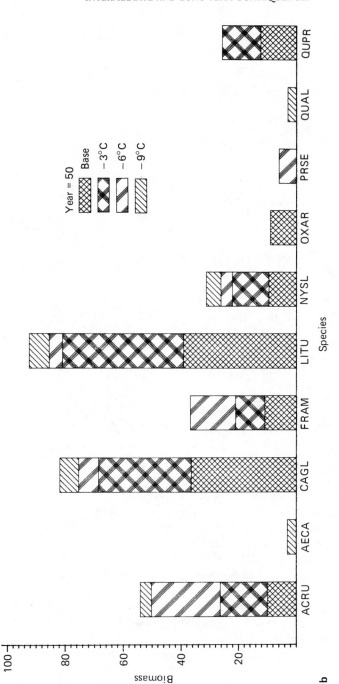

Figure 14. Effects of reduced air temperature on forest structure. Biomass values reported for species at 1 and 50 years (a and b, respectively), with temperature reductions occurring only for first 5 years. Species code: AECA, *Aesculus octandra*; ACRU, *Acer rubrum*; CAGL, *Carya glabra*; DIVI, *Diospyros virginiana*; FRAM, *Fraxinus americana*; LITU, *Liriodendron tulipifera*; NYSL, *Nyssa sylvatica*; OXAR, *Oxydendron aboreum*; PRSE, *Prunus serotina*; QUAL, *Quercus alba*; QUPR, *Quercus prinus*; SAAL, *Sassafras albidum*.

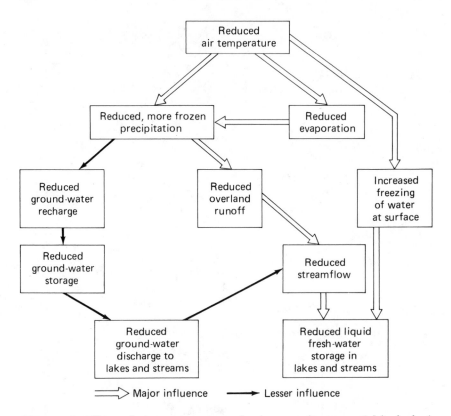

Figure 15. Effects of air temperature reductions on the terrestrial hydrologic cycle.

Lakes and reservoirs smaller than 100 km^2 are more important sources of water for most of humanity, being far more numerous, more widely distributed, and shallower than the larger water bodies. For the Earth as a whole, there are about 13 million of these smaller lakes, holding about 2000 cubic kilometers of water and averaging about 3 meters in depth (Bowen, 1982; Tamrazyan, 1974). These lakes' small volumes make them more susceptible to significant freezing, but possibly less significant as potential water sources. However, insofar as the surface area of lakes represents opportunity of access, these small lakes are extremely important, making up about two-thirds of the total lake surface area on the planet (Bowen, 1982; Tamrazyan, 1974).

Groundwater typically has a much longer residence time than water on the surface. Its warmer outflow would counteract some of the effects of air temperature drops. The reduction of groundwater recharge

Table 33. Selected Water Balance Terms of Northern Hemisphere Water Bodies.[a]

Budget term	Europe	Asia	North America
Rivers			
Average water stored (km^3)	80	565	250
Inputs (km^{3}yr^{-1})			
from overland runoff	2090	10660	5290
from ground-water	1120	3750	2160
Outputs (km^{3}yr^{-1})	3210	14410	7450
Lakes			
Average water stored (km^3)	2027	27782	25623
Reservoirs			
Average water stored (km^3)	422	1350	950

[a]Information summarized from USSR Committee for the International Hydrological Decade (1977, 1978).

resulting from the loss of liquid precipitation would cause outflow to slow if the reduction continued for an extended period of time. This effect would have a noticeable impact on rivers and lakes that are fed by shallow groundwater from glacial and alluvial deposits. Water nearest the surface of the Earth has the greatest chance of freezing because of atmospheric chilling. Given that lakes and reservoirs contain the bulk of the dependable stored water that might be accessed without pumping, it is appropriate to assess what portion of lake water storage would be vulnerable to freezing. Table 35 and Figure 16 report this for lakes in the Northern Hemisphere.

Table 34. Distribution of Fresh-water Lake and Reservoir Volumes and Areas.[a,b]

Area class (km^2)	Europe			Asia			North America		
	No.	Area (km^2)	Volume (km^3)	No.	Area (km^2)	Volume (km^3)	No.	Area (km^2)	Volume (km^3)
>10^4	1	17,700	908	4	92,670	23,200	8	327,280	24,322
10^3–10^4	26	74,989	995	21	67,070	3,128	22	73,185	1,258
10^2–10^3	23	9,618	479	36	16,760	520	17	7,252	214

[a]Information summarized from USSR Committee for the International Hydrological Decade (1977, 1978); Tamrazyan (1974); Bowen (1982).
[b]Table omits several larger lakes for which statistics were not available.

Table 35. Distribution of Lake Volume by Depth for Different Size Classes.[a]

Size class (km^2)	Fraction above					Share of hemisphere[b]	
	0.5 m	1.0 m	1.5 m	2.0 m	2.5 m	Lake area	Lake volume
>10^4	0.005	0.009	0.013	0.018	0.022	0.22	0.86
10^3–10^4	0.020	0.040	0.058	0.076	0.092	0.11	0.10
10^2–10^3	0.014	0.028	0.042	0.055	0.068	0.02	0.02
10–10^2	0.14	0.27	0.38	0.49	0.58	0.12	0.0076
1–10	0.28	0.50	0.68	0.81	0.91	0.19	0.0057
<1	0.56	0.88	1.00	1.00	1.00	0.35	0.0046

[a]Calculations were made assuming that lakes are shaped like elliptic sinusoids, whose volume is described by:

$$V = 1.456 \ abz$$

where a and b are radii of the surface ellipse and z is the maximum depth.
[b]Summarized from USSR Committee for the International Hydrologic Decade (1977) for lakes over 100 km^2 in area; from Tamrazyan (1974) and Bowen (1982) for smaller lakes.

Much research has been conducted into the thermal budgets of freezing water, particularly in Canada, the Scandinavian countries, and the Soviet Union. Since air temperatures would fall to below freezing levels rapidly after the nuclear war, ice would begin to form on most surface waters within a few days regardless of the season. Once initial ice has formed, its thickening can be predicted reasonably well as a function of the number of freezing degree-days. Ashton (1980) suggested the equation:

$$I = a(FDD)^{1/2} \tag{27}$$

where I = ice thickness
       a = constant
      FDD = cumulative freezing degree-days.

The coefficient a has a theoretical value of 3.41 (using centimeters and Celsius or Kelvin temperatures for units) based on idealized conditions. Its range in practice is about 1.7 to 3.07. Using this equation and empirical range, Figure 17 plots the potential ice thicknesses that could evolve at mid-latitudes for the baseline 5000 MT and 10,000 MT nuclear wars. Figure 18 translates these curves into estimates of the fraction of Northern Hemisphere fresh lake water frozen.

Ice thicknesses in middle northern latitudes could thus exceed one meter. Heat entering lakes with warmer groundwater would reduce this figure in some places. In the more widely accessible small lakes, even 0.5 meter of ice would tie up a major fraction of their volume. In the 5000

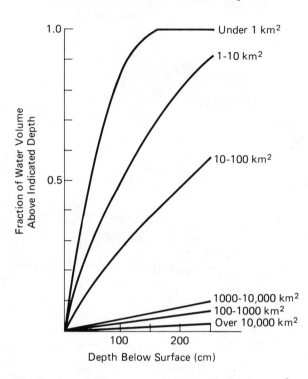

Figure 16. Distribution of lake water volumes with depth as a function of lake surface area. See notes to Table 35.

MT scenario, the shorter duration of the freeze would undoubtedly prevent some of the larger lakes from freezing over completely. Few lakes could remain open in the 10,000 MT scenario. Adding melting time and the normal climatic variation, many lakes could remain frozen over for a complete annual cycle. In addition to direct effects on freshwater biota, this situation could also lead to anaerobic conditions under the ice.

The water built up in snow and ice during the three-month or six-month sub-freezing periods would begin to melt afterwards, subjecting some areas to flooding. However, since the temperature would rise much more slowly than it fell just after the attacks, melting would occur at a relatively slower rate. Persons living in lake valleys for access to water would have to escape the floodplains.

During the period of frozen surface waters, human water supplies in much of the Northern Hemisphere would be restricted to groundwater and sub-ice surface water. Significant amounts of energy would have to

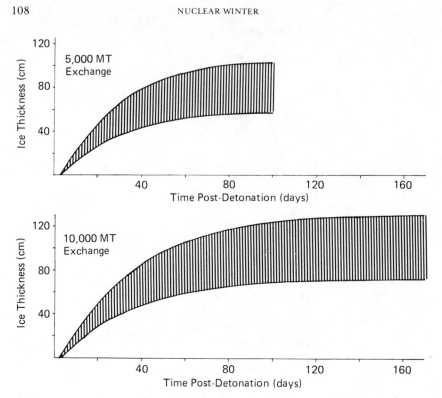

Figure 17. Projected thickness of ice on surface water systems from two nuclear war scenarios. Shaded areas indicate range of uncertainty.

be expended to melt snow, break through ice, or pump groundwater to access water for drinking and sanitation.

Freshwater organisms of the Temperate Zone would be subjected to a freezing regime longer and more extreme than they have evolved to fit. Their habitat would be reduced in volume significantly. The amount of light penetrating the water column would be greatly reduced because of the ice cover. Continuing oxygen demand from warmer bed sediments, reduced or eliminated oxygen exchange with the atmosphere, and drastically reduced photosynthesis in aquatic plants and algae could produce widespread "winterkill" resulting from anoxia (see Wetzel, 1975, Chapter 8).

Estuarine Ecosystems. Coastal aquatic ecosystems would receive less freshwater inflow than normal for an extended period, raising the salinity of estuaries. Estuaries would also directly experience stress from

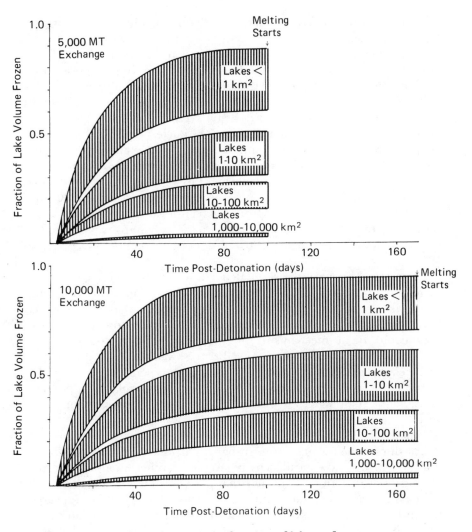

Figure 18. Extent of lake freezing as a function of lake surface area.

the reduced temperatures and light, but most estuarine systems would not be frozen to the degree freshwater systems would, because of exchange with warmer ocean waters.

To evaluate the effects of temperature reductions on estuaries, we reviewed the simulations performed on the Narragansett Bay Model, developed and analyzed by Kremer and Nixon (1978). Their model is a

multi-compartmental representation of the biotic, physical, and hydrologic components of the Bay ecosystem. As the case for the FORNUT
model, this estuarine model included both direct interactions of the
physical conditions with the condition of the constituent species and
indirect interactions that occur through inter-specific linkages.

No runs were made using this model for the present analyses. Rather,
previous parametric simulations done by Kremer and Nixon (1978) were
inspected for inferences about the effects of nuclear war. With respect to
temperature reductions, one simulation by those authors included
having the average annual temperature of the water (11.5°C) remain in
effect for each and every day during the year, effectively reducing
summer temperatures from the normal peak values of about 21°C and
increasing winter temperatures from the typical values of 3°C. The latter
is not reflective of the present analyses, but growth in the winter months
was not elevated that much compared to normal conditions (see Figure
19).

The original intent of the simulated run was to evaluate the influence
of temperature periodicity on the periodicity of phytoplankton and
zooplankton populations. Removal of temperature variations did in fact
have a profound effect on the dynamics of the plankton populations.
This could be a mechanism for substantial impacts on estuarine
ecosystems after a nuclear war, as well as freshwater and possibly coastal
and pelagic marine ecosystems, since normal periodicity would be
disrupted. This effect was more pronounced on zooplankton populations, where standing crops were reduced substantially for much of the
year.

That zooplankton effect could also involve the actual temperature
deficit during the warm season. In particular, the zooplankton populations normally show peak activity in the summer months, but the
Narragansett Bay Model indicates that period as the lowest in terms of
biomass under the constant temperature input. Phytoplankton appeared
not to be as affected, experiencing a somewhat earlier initiation of the
major bloom early in the year, a loss of periodic behavior thereafter, and
lower biomass during the normally high production period of mid-
summer.

From these simulations, we can infer substantial changes in the
planktonic dynamics in estuarine ecosystems resulting from changes in
water temperatures and in their seasonal patterns. However, the
probability of rapid recovery of such planktonic systems compared to
forested ecosystems suggest little long-term consequences from the first
year's extremes in air temperatures.

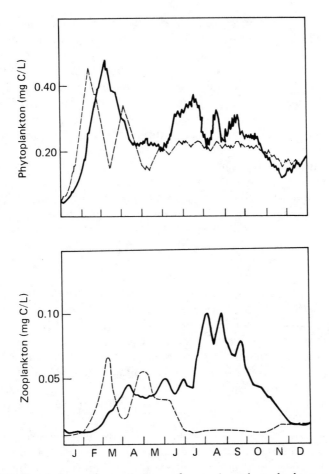

Figure 19. Simulations of populations of estuarine phytoplankton and zooplankton with normal (*solid line*) and constant (*dashed line*) water temperatures. (From Kremer and Nixon, 1978)

Effects of Reduced Light Levels

The projections from Turco et al. (1983a) suggest considerable reductions in the surface levels of incident sunlight for months after a nuclear war (see Figure 9). The potential for such hemispheric-level attenuation of sunlight to affect biological systems appears substantial. In this section, we will discuss some of the pertinent aspects of a light reduction stress.

Physiological Responses. Conversion of light energy into chemical energy via photosynthesis is the fundamental process driving natural and agricultural ecosystems. Disruption of photosynthesis by attenuation of incident sunlight would have consequences that cascade through foodchains that include humans as consumers, as well as other propagated impacts affecting biota upon which human survival may also depend. Primary productivity would be reduced roughly in proportion to the degree of light attenuation, assuming that the vegetation remained otherwise undamaged.

Many studies have examined effects on the rate of photosynthesis, plant growth, and crop yield. At least one researcher (Hutchinson, 1967) studied seedling mortality in response to prolonged periods of darkness. This study indicated differential sensitivity to prolonged darkness among terrestrial plant species, with those species whose habitats are usually in a forest understory having the greatest ability to withstand darkness. Reduced temperatures gave longer survival periods for seedlings in the dark because of lowered metabolic activity. However, tests were not conducted below 5°C, so the combined effects of extreme low temperatures and darkness were not studied. An interesting note is that for most of the species tested, longevity in darkness was 1–2 months, with death often associated with fungal growth. Most of the other research in this field has dealt with the influences on photosynthetic activity of the natural variations in light that now occur in natural habitats, but this provides some basis for assessing the possible effects of the reduced light levels following a nuclear war on the ability of plants to continue photosynthesis.

If leaves or suspensions of algal cells are exposed to increasing intensities of illumination, the CO_2 production occurring in darkness as a result of respiration is at first reduced and then replaced by net CO_2 uptake in photosynthesis (Figure 20). The point at which the photosynthetic activity just balances the respiratory activity of the organism is referred to as the compensation light intensity. At this point, the organism could maintain itself but could not grow. Without new growth there would be no new input of plant biomass into the foodchains. In fact, continued consumption by herbivores would most likely cause defoliation of higher plants and large reductions in the populations of algae if light were at or below the compensation intensity.

As the intensity of illumination is increased further, there is a region where the uptake of CO_2 increases proportionally with increases in illumination (Figure 20). The slope of this line reflects the efficiency of the photosynthetic process and is remarkably constant among plants. About 10% of the total energy of each additional increment of sunlight is

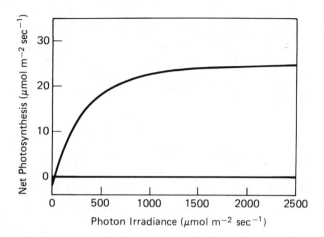

Figure 20. The rate of net photosynthesis by a single leaf of *Atriplex triangularis* as a function of the intensity of sunlight falling on the leaf. A leaf perpendicular to the solar beam would receive about 2,400 μmol m^{-2}sec^{-1} photons (400–700 nm) at noon on a cloudless day. In the dark the leaf respired CO_2 (net photosynthesis was negative); the compensation intensity (zero net photosynthesis) was reached at 50 μmol m^{-2}sec^{-1} photons, and the rate of net photosynthesis became saturated at about 1,000 μmol m^{-2}sec^{-1} photons. (Based on information in Björkman, 1981)

stored in the form of chemical bonds (for example, production of glucose from CO_2). The net efficiency is, of course, lower because some light is needed to offset respiration. Finally, as the intensity of illumination is increased still further, the marginal efficiency begins to fall and the process saturates, such that no further increases in photosynthesis occur with increases in light.

In general, the amount of light required for plants to reach compensation or saturation is proportional to the intensity of light normally absorbed in the native habitat. As a result, very little light is wasted. This can be illustrated by considering a forest canopy (Figure 21). The leaves of the trees that form the top layers of the canopy have a high light requirement for photosynthesis. If each leaf were perpendicular to the sun on a bright day, these would have reached rate saturation with light intensity, but since the leaves are usually at an angle to the sun, the quantity of light they actually absorb is less than saturating. Other leaves deeper in the canopy receive light entering through gaps, reflecting off the surfaces of or passing through the top layers of leaves. As the light intensity is reduced with depth in the canopy, the characteristics of the

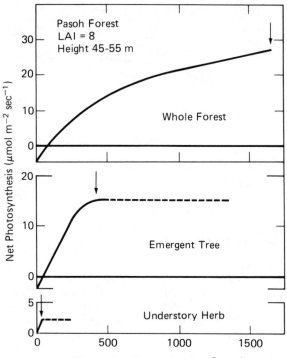

Figure 21. The rate of net photosynthesis of the whole canopy of a tropical rain forest expressed per unit ground area, and the rate of net photosynthesis per unit leaf area of individual leaves of an understory herb and an emergent tree in that forest. The arrows indicate the maximum photon irradiance reaching the top of the forest canopy and the native habitats of each of the leaves on a clear day. The leaves are shaded and receive only a fraction of the incident light; hence it requires a higher intensity to saturate the whole canopy than it would a single leaf within it. Reducing the photon irradiance at the top of the canopy from that indicated (arrow) would reduce the photosynthesis of all leaves within the canopy. (From Mooney et al., 1984)

plants change such that they require less light for compensation and saturation.

In very complex ecosystems such as tropical rainforests, different species have become specialized to occupy particular regions along this light gradient. Emergent trees and understory shrubs thus have quite different light response characteristics, even though they occur in the same forest.

The light response characteristics of the whole forest canopy is a complex sum of the light absorbed by each of the components of the canopy and the respective responses of those components to light. The net result is that it takes much higher light intensity to saturate the whole canopy than it does to saturate any single component of it. If the system is not saturated, then any attenuation of light will result in an attenuation of photosynthesis.

Dense stands of agricultural crops respond much the same as the forest canopy, showing a linear increase in photosynthesis per unit ground area with increasing light intensity (Figure 22). The data plotted in this figure are actual measurements of gross photosynthesis of a stand of cotton plants from the lowest to the highest intensities over the course of several typical days. Similar data exist for many other species. Over the course of a normal day, mature plants respire over the night and during

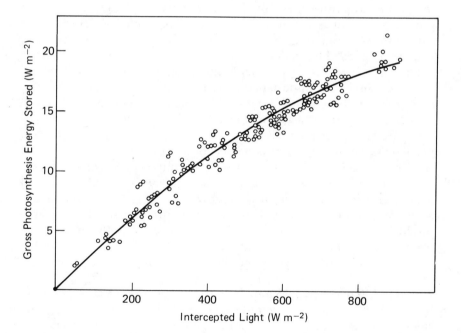

Figure 22. Gross photosynthesis (obtained by subtracting respiration) of a stand of cotton plants as a function of the light intercepted by the entire canopy. Photosynthesis is expressed as the heat equivalent of the products formed, and light is expressed as the heat equivalent of the light intercepted. The energy storage by photosynthesis increased linearly up to the highest irradiance levels reached on typical clear days. The overall efficiency ($\sim$3.3%) is similar to that of single leaves. (Redrawn from Baker et al., 1972)

the early morning and evening when the light intensity is below the compensation point. Because of these nocturnal losses, it takes a large fraction of the photosynthesis of a day for the crop to compensate for the respiratory losses. Respiratory losses as a fraction of gross photosynthesis range from about 30% for a wheat crop to about 75% for a tropical rainforest. Any reduction of light intensity would result in an attenuation of net crop photosynthesis, and if the total daily light dose fell below this maintenance level, the crop would cease all growth. Many studies have examined the effects of shading on the rate of photosynthesis, plant growth, and crop yield. Photosynthesis is usually reduced with any shading, as is total growth [which after all is one of the most direct measures of photosynthesis (Björkman, 1981)]. Crop yield is not always similarly affected (Christy and Porter, 1982; Zelitch, 1982) because some crop plants are not able fully to convert marginal increases of photosynthesis into a harvestible form. This is a significant consideration with respect to crop improvement, but at the levels of light attenuation projected here, it is highly unlikely that crops could even survive, let alone produce a harvest.

In discussions of limitations on plant growth, it is often stated that the availability of water is the primary limiting factor. If this is true, how can we attach so much importance to reductions in light intensity? The answer is that these factors limit in different ways. As explained above, light is an important factor in controlling the productivity of leaves or canopies. Water has little direct effect at this level, provided it is present in adequate supply. The primary effect of water limitation is to limit the amount of vegetation, and light incident on unvegetated areas is useless for photosynthesis. On a geographic scale, water limits the area and density of fields, forests, shrubs, and grasslands. Given constant light, it is the extent of this photosynthetically active surface that determines the productivity of an area. But if light intensity is reduced, the productivity per unit area would be reduced, also yielding a loss in productivity. Thus, the limitations by water and light are not mutually exclusive.

Grassland Ecosystems. Extrapolations from physiological responses to light reductions to ecosystem-level effects are based on computer simulations, as with the analyses of reduced temperature effects. The same models were run or analyzed (SPUR, FORNUT, Narragansett Bay Model) based on the same types of control analyses (see discussion on temperature effects for details).

Results from the grassland simulations (Table 32B) indicate about the same level of reductions in productivity for the insolation reductions analyzed as for the temperature reductions. Since the projected light levels from the second and later years after a nuclear war are not as reduced as the greater part of the range analyzed here, more long-term consequences on grasslands (and by inference crop monoculture systems) would ensue from temperature reductions than from light reductions.

Simulations were not run for the first few months after the nuclear war. As before, conditions with respect to light and temperatures are considered so extreme that the model would neither be accurate (parameters far beyond appropriate bounds) nor necessary; i.e., the response of the primary productivity of grassland ecosystems during the maximum temperature and light reductions is clearly to shut down all photosynthetic activity and to abandon the aboveground living biomass.

Forested Ecosystems. The FORNUT model was used to examine the effects of light reduction on the forest structure and productivity for eastern deciduous forests. Simulations included reduction in incident light by 50% for the first growing season after a nuclear war. Reductions in air temperature were not included in these simulations. Results are shown in Figure 23a and b.

Simulations of the effect of light reductions on forest growth suggest that productivity may be diminished over a 50-year period. However, the reduction in growth did not occur in the simulations immediately following the year of light reductions but was delayed some twenty years following the event. This likely results from the suppression of understory trees during the year of 50% light reduction. Under these conditions, light availability beneath the canopy would be very low. Since understory tree growth does not normally represent a large proportion of the total growth of the forest, this suppression was not immediately obvious in the total stand biomass values. However, by year 20 the overstory has experienced some loss of individuals through natural mortality. At this point, the growth of understory trees into these gaps in the canopy is important for the maintenance of stand productivity. In the simulation with light reduction, understory individuals had not regained their productive potential and were not as capable of fully utilizing gaps in the canopy as they were in the base case simulation. Consequently, the simulations show an unanticipated delay in the onset of a reduction in forest productivity rates.

These results are not offered as the definitive projection of the changes to occur in a deciduous forest subject to a period of light reduction. Differential sensitivities to reduced light among groups of tree species could differ in specifics from the values in the model, and effects from

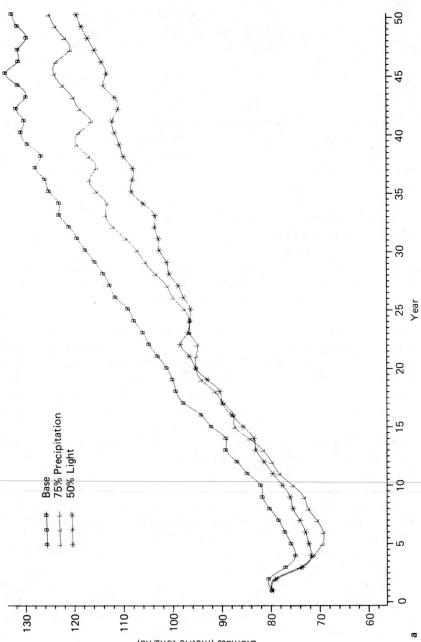

a

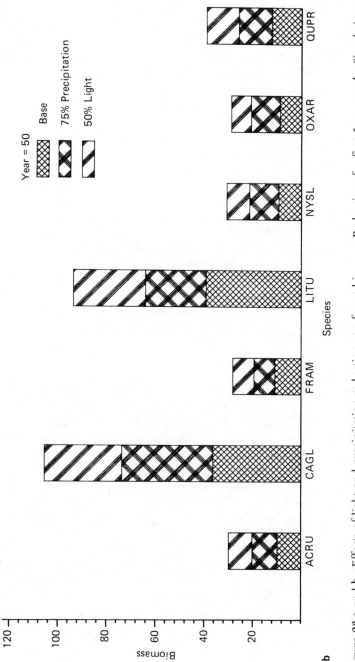

Figure 23 a and **b**. Effects of light and precipitation reductions on forest biomass. Reductions for first 5 years only. Simulations performed using FORNUT model. Species code: AECA, *Aesculus octandra*; ACRU, *Acer rubrum*; CAGL, *Carya glabra*; DIVI, *Diospyros virginiana*; FRAM, *Fraxinus americana*; LITU, *Liriodendron tulipifera*; NYSL, *Nyssa sylvatica*; OXAR, *Oxydendron aboreum*; PRSE, *Prunus serotina*; QUAL, *Quercus alba*; QUPR, *Quercus prinus*; SAAL, *Sassafras albidum*.

temperature and precipitation could confound the results from decreased light levels. However, the model results do illustrate the potential for long-term consequences (over decades) from relatively short-term conditions (over months), including consequences that do not even first appear until many years after the system is stressed.

Estuarine Ecosystems. The effect of light reduction on estuarine and marine ecosystems could be much more extensive and significant than effects from changes in air temperatures. In part, this is because the water temperatures for those ecosystems are relatively buffered from the extremes experienced by terrestrial or freshwater ecosystems. Also, estuarine and marine ecosystems tend to have light limitations with depth, so reduction in solar insolation would reduce the depth to compensation points, perhaps to the point where no net photosynthesis could occur in the water column.

Again, the simulations performed by Kremer and Nixon (1978) are used for inferences about light reduction effects on estuaries. The simulation runs were performed to evaluate the effects of removing annual insolation periodicity on the periodicity of the plankton community. However, here different levels were chosen for the constant daily solar input values, ranging from 25 ly/day to 200 ly/day.

Results indicated that constant light levels at or above normal average levels allowed the periodicity to remain, with very little effect on phytoplankton dynamics (Figure 24). However, additional light above the low values normally experienced in winter allowed phytoplankton blooms to occur much sooner and at greater levels than normal, suggesting light-limiting conditions normally occur during those months. Conversely, low light levels, when reduced by a factor of 4 or so, do result in drastic alterations in biomass for populations of algae and zooplankton. This suggests that markedly reduced levels of solar inputs during the first few months could cause widespread plankton population crashes in estuarine ecosystems and, by inference, at least as dramatic changes in marine phytoplankton productivity. However, planktonic forms could survive such population crashes, with enough encysted individuals to allow rapid population recovery when light levels resume more typical values. How a population crash for phytoplankton would affect higher trophic levels is difficult to predict. In general, those organisms whose primary source of energy is plankton-based could suffer significant losses if the plankton population levels were sufficiently low for an extended period of time, i.e., beyond the duration of energy reserves. Those fish and other marine organisms that rely primarily on a

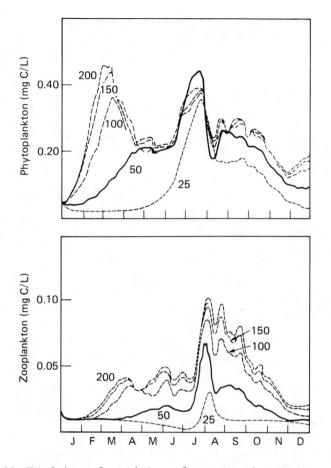

Figure 24. Simulations of populations of estuarine phytoplankton and zooplankton with normal (*solid line*) and seasonally constant (*dashed lines*) light levels at 25, 50, 100, 150, and 200 ly/day. (From Kremer and Nixon, 1978)

detritus base would not likely experience adverse effects from the projected reductions in light.

Effects on Agricultural Production

The effects on agroecosystems would reflect the effects on other biologic systems. However, since agroecosystems are so central to the maintenance of current levels of human populations, the importance of this type of system demands much closer attention be paid to it. This section provides an overview of the factors affecting agricultural production.

Food, along with water and shelter, is essential for human life. The agricultural systems of the world provide society with adequate quantities of nutritious plant and animal foods.

In the United States each person on average consumes almost 1,400 lb of food annually, including 218 lb of meat; 308 lb of dairy products; 60 lb of fats; 161 lb of fruits; 308 lb of vegetables; 150 lb of grain; 133 lb of sugar; and 12 lb of other food (USDA, 1981). Most of these foods are produced by U.S. agriculture. In addition to supplying domestic needs, the United States is also the largest food exporter in the world.

Agricultural production depends on several natural resources, including fertile land, water, appropriate temperature conditions, sunlight, fossil energy, and human labor. All are essential and must be supplied in adequate amounts or within suitable ranges for effective crop production. Alter temperature, rainfall, or any other basic resource, and crop yields are severely reduced. The effects of such changes have been demonstrated in U.S. crop production during the past decade by droughts and unseasonal freezes (USDA, 1981).

This section discusses the effects that nuclear war-altered sunlight, temperature, and rainfall conditions (Turco et al., 1983) could have on U.S. agricultural production and subsequently on human food supplies.

As we have seen, the impact of a nuclear war would extend well beyond the devastation caused in the target areas. Sunlight could be decreased in the Northern Hemisphere by 99% (Turco et al., 1983a). This would have a three-fold effect. First, the solar energy needed for plant photosynthesis would be severely curtailed. Second, with decreased sunlight the temperature could decrease to as low as −25°C (Figure 8) (Turco et al., 1983a). Combined, the decreased sunlight and lower temperatures could reduce rainfall because of the change in available energy needed for the hydrologic cycle. Further, most surface water would be frozen and thus unavailable for easy use by humans, crops, and livestock, as discussed previously.

The principal crops produced in the United States are wheat, corn, rye, barley, sorghum, soybeans, dry beans, peanuts, potatoes, tomatoes, peas, fresh beans, carrots, sweet corn, lettuce, cabbage, broccoli, citrus, apples, plums, peaches, and pears (USDA, 1981). Except for the tree fruits, most of the crop plants are tropical annual plants. Together these plants produce about 95% of the food energy in the United States. Although the grains dominate this supply in the food system, 90% of the grains used in this nation are fed directly to livestock (Pimentel et al., 1980a).

Tropical food crops are grown in the temperate region of the United States because the region has a 3 to 4 month summer period that is ideally suited for culture of these short season crops. In addition to their fine nutritional qualities, tropical annuals have the advantage of high yield during the short season of a U.S. summer when growing conditions are favorable. Also these annual crops, especially the grains, are relatively resistant to attack from pest insects, pathogens, and weeds.

Most tropical crops require daytime temperatures of 20°C or higher for growth and fruit formation. The exposure of crops even to low, non-freezing temperatures for short periods during the growing season will severely reduce yields in some crops. Rice and sorghum, for example, when exposed to 13°C, especially during grain formation, will have significantly lower yields (Larcher and Bauer, 1981). Low temperature causes many plants to produce sterile pollen. Major crops like corn and soybeans also are quite sensitive to low, non-freezing temperatures in the range of 10°C (Larcher and Bauer, 1981).

Reductions in the growing season temperatures by 1°C can decrease the yields of corn by 440 kg/ha, or 7% of normal yields of 6270 kg/ha (Johnson, 1983). Also, reducing the mean temperature by 0.6°C may shorten the growing season and time between killing frosts by about 2 weeks (Malone, 1974). Reduction in the growing season of corn, for example, by 2 weeks may decrease yields by 10 to 15% in the United States (Pimentel and Pimentel, 1979). In Canada, where normal growing seasons are shorter, reducing the length of the growing season by 10–20 days could totally eliminate wheat production, as suggested by recent computer simulations performed by Agriculture Canada (Harwell and Stewart, in prep.). Such a situation would ensue from only a 1–3°C reduction from normal growing season temperatures.

Although some crop plants can tolerate frosts and freezing temperatures, these same crop plants are sensitive if they have not been acclimatized to low temperatures over a period of time. Winter wheat, for example, can tolerate low temperatures in the range of −15°C to −20°C when conditioned slowly, as occurs during the fall months (Levitt, 1980). However, wheat plants are killed by only −5°C if they are exposed during summer growth.

Responding similarly to wheat, cabbage can tolerate temperatures in early spring or late fall as low as −10°C. However, when taken from greenhouses kept at 26°C, they are easily killed by a frost and exposure to only −1°C (Pimentel, personal communication). Thus, some crop plants might survive the projected low temperatures if they could be slowly conditioned to low temperatures. However, they would not be able to survive as temperatures declined quickly following a nuclear war (Turco et al., 1983a).

Insolation. Crop plants are also sensitive to changes in sunlight intensity and day length (Björkman, 1981). Reducing the light levels by one-half on individual leaves usually does not affect the growth and yield of crops (Björkman, 1981). However, most plants under normal growth conditions have several layers of leaves, many of which are completely shaded. Therefore, while a 50% or less reduction in light might not decrease photosynthesis and growth of a plant if all of its leaves were exposed equally to the sunlight, a small reduction of 10% in ambient light levels might decrease normal growth of the crop and yields (Björkman, 1981). The projected 99% reduction in sunlight would prevent most plants from growing and cause them to deplete any stored reserves of nutrients. Thus, crop production would not be possible if sunlight were reduced 99%, and most crop production would be impossible if sunlight levels were reduced even by 50%.

Moisture. After sunlight, water is the most important factor limiting U.S. crop plant growth and production. Crops require and transpire massive amounts of water; for example, a corn crop that produces 5600 kg of grain per hectare will take up and transpire about 2.4 million liters of water per hectare during the growing season (Penman, 1970).

About 12% of U.S. crop production is irrigated (USDA, 1980), and most of this is located in the West and Southwest. Agricultural irrigation consumes 83% of the total of 360 bld (billion liters per day) that is consumed (Figure 25). Irrigation would be very difficult during the early years after a nuclear war. Surface waters initially would be frozen, and most groundwater sources would be impossible to tap without fuel and pumps, neither of which would likely be widely available.

For normal rain-fed crop production, rainfall levels determine what crops can be grown in a given area. For example, with 75 to 125 cm of rain annually, corn, rice, and beans can be grown; with 50 to 75 cm, wheat, oats, and rye; and with 25 to 50 cm of rain, sorghum and millet could be grown. Thus, if rainfall were reduced by one-half or more in the aftermath of a nuclear war, normal crop production would be impossible in most agricultural regions.

Post-Nuclear War Crop Yields. The key question, of course, is how much food could be produced after a nuclear war. The answer is simply

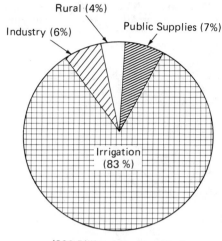

Rural (4%)

Industry (6%)   Public Supplies (7%)

Irrigation
(83 %)

(360 Billion Liters per Day)

Figure 25. The distribution of freshwater consumption in the United States. (From Murray and Reeves, 1972)

very little, if any (Table 36). The extent of devastation of agricultural systems would depend on the time of year that a nuclear war took place. If the nuclear war occurred during the winter, then conceivably some of the winter wheat and other crops that were already in the ground might be able to survive and grow in the altered summer that would follow. Sunlight and temperature conditions, however, during the summer immediately after the war would be extremely unfavorable in most locations. The only exception would be temperature conditions near large bodies of water, which would be moderated to some degree; however, here one could expect particularly violent weather because of strong lateral temperature gradients. Thus, it is unclear how advantageous locally moderated temperatures would be.

Projected crop production after a winter nuclear war might be at best 1–10% of normal, considering only the effects of reduced temperatures and light (i.e., not accounting for possible decreases in precipitation or for the certain reduction in societal support to agroecosystems; such effects would significantly reduce productivity further). But if the attack occurred during the summer and growing conditions were altered as predicted, then crop yields would be essentially zero for that growing season (Table 36).

Considered together, most of the livestock species are more tolerant of climatic changes than crop plants, assuming their food resources are not

Table 36. Current Production of Various Crops and Animal Products and Projected Production for the Year After a Nuclear War (million metric tons).

		Projected production[b]	
	Current production[a]	Winter war	Summer war
Meat products	28	10	5
Dairy products	32	10	5
Eggs	4	0	0
Grains	307	20	0
Vegetables	32	3	0
Fruits	17	2	0
Sugar	14	0	0
Total	434	45	10

[a]Summarized from USDA (1981).
[b]Assumes survival of one-third of livestock and some survival of winter wheat after a nuclear war occurring in the winter, after the wheat had become hardened. The production estimates do not take into account the effects on agriculture of loss of societal support, such as lack of fertilizers, pesticides, human and mechanical labor, etc. Thus, these values are considered to be overestimates.

destroyed. Many cattle, sheep, and goat breeds can survive relatively low temperatures, especially if provided with minimal amounts of shelter and food (NAS, 1981). The few beef and dairy cattle and sheep that survived the initial nuclear event might continue to survive on the remaining forage. However, livestock numbers and productivity would be greatly reduced because of inadequate supply of feed grain (Pimentel et al., 1980a). Optimistically, perhaps as many as one-third might survive a winter attack and be kept in production (Table 36). However, a summer attack would destroy most of the forage and leave the livestock severely stressed for food.

Most young chickens and swine husbanded in this country can not tolerate temperatures below 14°C (NAS, 1981). Thus, the reduction of temperatures to −25°C would seriously impair poultry and swine production without substantial protection by human-provided shelter and heat. Further, the lack of essential grain as food for these animals would curtail their productivity.

Equally important as the altered climatic conditions for crops and livestock is the lack of equipment, fuel, fertilizers, pesticides, some feeds, and human labor for crop and livestock production. Industrial production facilities that are located near large cities would be either destroyed or shut down. In addition, the agricultural transport system would be inoperative. Thus, agricultural production would be without needed fuel, fertilizers, pesticides, and some seeds, in particular the high yielding hybrid seeds. Without fossil fuel for farm machinery, there would be no way, except for manual labor, to till and plant, because there are few draft animals to provide power, and fewer still could be expected to survive into the post-nuclear winter period.

Even now large-scale crop production based solely on human labor is impossible in this country, because of the tremendous labor requirement. About 1200 hours of labor are required to raise a hectare of corn by hand (Pimentel and Pimentel, 1979), while in the United States today only 12 hours of farm labor is required (Pimentel and Burgess, 1980). Thus, raising corn by hand would require a 100-fold increase in the farm labor force, and this would be impossible after a nuclear war.

Shortages of fertilizers and pesticides would also reduce potential yields, but the primary problem would be the impossibility of tilling and planting vast acreages without tractors and fossil fuel energy (Pimentel, 1980). In addition, over one-fourth of the agricultural land would probably be contaminated with radiation such that it could not be used for crop and livestock production for at least the first year. The agricultural land that surrounds most urban areas and military bases would be most likely contaminated with high levels of radiation.

Although most of the agricultural and food production system would be destroyed following a nuclear war, the United States is fortunate in having large quantities of grains available for livestock production and export. Immediately after a nuclear war, export of grains [120 million tons per year (USDA, 1981)] would be terminated because of the inoperative transport system. In addition, most grains would probably not be fed to livestock but would be conserved for human food.

Some grain stores would be destroyed by the war, but conceivably substantial quantities could be left intact, sufficient grain to keep the present U.S. population eating as vegetarians for six months to three years, assuming some legumes such as soybeans would also be available. The difficulties would be in distributing the grain to the survivors and protecting it while in storage. Although the grains (including soybeans and other legumes) might be adequate to supply suitable amounts of calories and proteins, the population would be malnourished in terms of vitamins B_{12}, C, A, riboflavin, calcium, and iron. Vitamin D, because of lack of sunshine and few food sources, would also be a growing concern as body stores were used up.

Three-fourths of the U.S. population live in urban areas (U.S. Bureau of the Census, 1981). These people keep on hand only 2–7 days supply of food and depend on nearby grocery stores that, on average, keep little more than a week's supply of food items in stock. Thus, within one week the people of the major cities would be without adequate food, especially because of the inoperative transport system. The large amounts of grains located in rural areas would be unavailable to people in the cities.

An efficient transport system not only is essential for the distribution of food to consumers, but also functions to bring needed production supplies to farmers. On average about 600 kg of goods and supplies must be transported to farms for each hectare that is cultivated (Pimentel and Pimentel, 1979). Available data indicate that about 60% of the goods are transported from factory to farm by rail and the remaining 40% by truck. The average distance that goods are transported to deliver them to farms is about 640 km (Pimentel and Pimentel, 1979).

Annually in the United States, about 160 million hectares of cropland are harvested, and an average of 3400 kg of food and feed products are harvested per hectare. In contrast to farm supplies, the distribution of harvested food products to consumers is carried out primarily by truck (60%) (Pimentel and Pimentel, 1979). The remaining 40% is transported by rail. On average the distance of moving farm products is also about 640 km.

The transport of fruits and vegetables provides quite a different picture from grains and other foods. Most fruits and vegetables are now produced in California and Florida (USDA, 1981) and transported an average of 3000 km to consumer markets.

Agricultural production also depends upon the natural biota to keep the agroecosystem functioning by recycling organic wastes, fixing nitrogen, and controlling pests (Pimentel et al., 1980b). Most of the natural biota essential to agriculture exist in the litter and soil. In the short term, these probably would survive the projected low temperatures. If their source of plant matter were reduced for a long time period after a nuclear war, then their numbers could decrease as well as the beneficial functions they perform.

Several insects, especially honey and wild bees, are essential for the pollination of 90 U.S. crops, currently worth about $4 billion (Pimentel et al., 1980b). Thus, if the number of these pollinators were reduced and/or eliminated because of low temperatures and insufficient sunlight, production of many crops would be severely reduced. Unfortunately, insect pollination of crops is not a technology that can be replaced by humans or machine. The magnitude of work done by just the bees (Pimentel et al., 1980b) is illustrated by the fact that it is calculated that on a sunny summer day in New York State, bees pollinate over 1 trillion (1×10^{12}) blossoms, making possible fruit formation in many crops.

A number of previous studies have examined the consequences of nuclear war on agricultural production in the United States. Brown et al. (1973) typified those studies that focused on fallout effects on crops and livestock animals. Brown and Pilz (1969), Haaland et al. (1976), and Billheimer and Simpson (1978) evaluated effects of nuclear detonations on the transportation, storage, and distribution of agricultural products to a relocated population of survivors. Essential consensus was reached among those authors, each sponsored by the U.S. Defense Civil Preparedness Agency, and the authors of a briefing report by the Federal Emergency Management Agency (FEMA, 1982) that while food shortages might become important locally, in general the effects of a large-scale nuclear war on agriculture and food availability would not significantly affect human survivors. We clearly come to quite different conclusions, that the nuclear winter alone could result in essential termination of the agricultural production and distribution system and that widespread starvation could ensue. Further, even without the effects of a nuclear winter, disruption in food production and distribution have been analyzed by Hjort (1982) to lead to insufficient availability in a few months and to subsequent severe food problems. If a nuclear war results in reducing average air temperatures to −20°C or below, sunlight by 99%, or rainfall by a half or more, agriculture and the associated food production of plants and animals in the United States would be practically impossible the first year. This would be especially true if the war were initiated during the summer growing season. All of the projected climatic changes would have a much greater effect on agricultural production than the initial blast, fires, and radiation. The effects of the blast, fire, and radiation from the nuclear explosions could destroy about 10% of all crops and livestock, but it is the reduced temperature, sunlight, and rainfall conditions that would seriously impair agricultural production.

Assuming that the grain in storage were conserved and not fed to the surviving livestock, grain and legume supplies would be potentially large enough to provide an adequate vegetarian diet of protein and calories for the U.S. population for one-half to three years without further grain production. However, without fruits and vegetables and some animal products, serious malnutrition would result. In particular the population would be malnourished in terms of vitamins A, B_{12}, C, and riboflavin, and to a lesser extent calcium and iron.

However, although there theoretically might be adequate grains for the surviving population, there would be no means to transport and distribute the available food to the U.S. population, which resides mostly in urban areas. Thus, stored grains would be available only to the rural population (currently about one-quarter of the U.S. population). Therefore, even though 50–100 million Americans might survive the initial effects of blast and radiation, food shortages during the subsequent years would result in a famine for many millions in the United States.

The propagation of this effect beyond the United States and other targeted countries would cause enormous consequences for the rest of the world. The cessation of food exports from North America alone could seriously threaten the food sources of many nations. Further, typically those food importers have little storage of grains and other foods (USDA, 1982; FAO, 1983), and effects on these marginally nourished peoples could occur rapidly. Indeed, recent analyses indicate the total grain storage in the world is only 40 days' worth of world consumption, down from 100 days in 1960 (Brown, 1981). Adding to this, of course, is the effect of the nuclear winter on targeted and nontargeted countries alike in the Northern Hemisphere and, probably, the Southern Hemisphere as well. Reductions in light, temperature, precipitation, and chemical subsidies could be expected to curtail severely or to eliminate agricultural productivity throughout the world. An important factor is that current grain production involves the utilization of land of marginal productivity, which is therefore especially sensitive to only relatively minor climatic imbalances (Barr, 1981). It may well be that the greatest effect on humans from a large-scale nuclear war would be famine. Globally, on the order of 10^9 people could die from starvation.

Societal Disruptions

As part of the presentation and analysis of the long-term consequences of nuclear war, the impacts of the nuclear detonations on the human ecology must be examined. The collective effects of the war on human social systems are themselves of extreme importance, insofar as those who survive would determine recovery processes during the post-war period. A synergistic effect could be expected to result, with ecosystem-level processes affecting the survivability of humans, and human social systems potentially mitigating some of the fatality levels produced by those ecosystem effects.

A thorough treatment of social effects would anticipate nuclear war-induced changes in such diverse social arenas as food production and organization, human services provision, housing and infrastructure maintenance, energy generation and economic exchange, transportation and urban rehabilitation, and disaster relief and institutional recovery. Little anticipatory analysis is available on these topics, with the important exception of the medical catastrophe occassioned by disappearance of medical expertise and health care systems in detonation zones.

The time frames under discussion here extend from the first few days immediately succeeding the detonations posited in the initial conditions analyses through the following decade. While there are no close parallels to be drawn from human history that could adequately illustrate the potential societal effects of such a nuclear war, some conclusions can be drawn from past experience that can assist in bounding the potentialities of human reactions. Operating within these bounds, the following discussion will examine, in a stepwise manner, the potential pathways of societal responses to a large-scale nuclear war, focusing on social systems demonstrated in the highly industrialized nations.

Immediate Post-War Effects. From the moment of detonation of the nuclear warheads and through the following several weeks, severe physical and psychological problems would confront those who have survived the immediate blasts. While the long-term effects of the remaining, functioning social systems would be minimally felt during this period of extreme disorder, some degree of survivability would depend on those systems. Following is an examination of the potential impacts during this immediate post-war period dealing both with social systems as entities and with individual ranges of responses.

Medical and Emergency Response. Within minutes of the nuclear war, analyses here show that 125–170 million people in the United States alone [and one billion people worldwide (Bergstrom et al., 1983)] can be expected to die from blast and other direct effects. Severe medical care problems would result because of the additional large numbers of people [30–50 million in the U.S., one billion worldwide (Bergstrom et al., 1983)] who would require primary medical care for burns, radiation sickness, blast-related physical traumas, and shock. The concentration of medical facilities, trained personnel, and medical supply sources in targeted areas would result in massive losses of medical resources at a time of unprecedented demand. For example, 71% of physicians and 57% of hospital beds are located in the urban areas of the United States. The pharmaceutical industry may be one of the critical industries targeted for maximum destruction (Katz, 1979). Lack of access to surviving facilities, both for victims and those trained to aid them, is highly probable because of the blocking of transportation routes by rubble, destruction of transport mechanisms, and inability of survivors to render aid to those injured.

Deaths from injuries would be increased by the inevitability of multiple injuries to many victims, such as combinations of burns, physical traumas from blast, resultant shock, and radiation injuries. Hiroshima data indicate nearly 40% of survivors sustained multiple injuries (Katz, 1982). Radiation exposure would increase susceptibility to infection, especially through the known synergy between burns and radiation which profoundly increases mortality rates (Abrams and Van Kaenel, 1981). The destruction of concentrated sources of medical supplies, particularly antibiotics and drugs for chronically ill persons, would further increase deaths in the time period immediately following detonation.

Emergency responses would be limited by the lack of surviving personnel responding to assigned work stations, by destruction of vehicles and lack of transport routing access, and by interrupted communications. Individual efforts would likely be limited through psychological shock and through the higher importance placed on assuring personal and family member survival than on altruistic aid to victims, exacerbated by the overwhelming desire to flee areas of great destruction, as was illustrated at Hiroshima (U.S. Strategic Bombing Survey, 1946; Katz, 1982, citing Siemes, 1946; Siemes was a Jesuit professor who was present in Hiroshima at the time of the nuclear bombing). These efforts would be further complicated by the widespread occurrence of flash blindness (Glasstone and Dolan, 1977), which would last over the immediate time after blast, resulting in a large, temporarily sight-disabled population just at the time when maneuverability would be most important.

Societal Response. During the immediate post-war period, social systems would largely be chaotic. Victims of immediate blast death and disabling blast-related injuries would include proportionate numbers of police, fire, sanitation, utilities, and local government personnel and facilities, and disproprotionately higher numbers of military personnel and installations. These are the traditional respondents to crises, and their unavailability would increase the probability of death for those severely injured and those trapped in death-threatening circumstances. Individual aid and rescue efforts could not, even under ideal circumstances, replace in any real capacity the assistance available during more ordinary crises. Social order would not be maintainable through single-person actions, and even lay group actions would be unlikely to coalesce during this immediate time frame. Fires would likely spread unchecked by human intervention, and interruption of water supplies and utilities services would create further hazards for those surviving the immediate blast effects. Electromagnetic pulse produced by the nuclear detonations would damage communications equipment and transportation systems and would severely disrupt electrical power systems.

There would be intense competition almost immediately for basic sustenance requirements such as shelter, clothing, food, and water. Local governments would be unable at this time to handle the normal functions of distribution of supplies, and survivors would be unable to assess for themselves the extent of devastation and therefore the probability of receiving outside aid. If there were concentration of uncontaminated food and water at particular sites after the initial detonations, communication of this information to needy individuals would be extremely difficult for several weeks or longer.

The functioning of systematic survival efforts at this post-war stage would likely focus on efforts made by small groups of survivors, primarily family units and those grouped together by circumstance at the time of the blast. The experiences in Hiroshima and Nagasaki suggest that the immediate exodus of people fleeing the target area would reverse within a day to a tremendous influx of people in search of relatives and friends (U.S. Strategic Bombing Survey, 1946). Lack of communication of information about other surviving groups or individuals, especially family and friends, would enhance isolation of groups and individuals and lessen the effectiveness of resource pooling for enhanced survival. Those seeking pathways of escape from the areas of central destruction would be in competition for portable supplies with those persons who choose to stay nearer urbanized areas, thus increasing the difficulty of maintaining concentrated sources of essential materials and increasing, as well, the pressure to seek personal or very small group survival as opposed to societal survival.

Individual Responses. At the individual level, two interactive responses to the destruction would influence survivability, not only of the affected individual but also of the collective individuals who would make up the survivor pool groups entering the intermediate post-war time range. These responses are at the physical level and at the psychological level, and without doubt would profoundly act together on individuals.

Physically, human survivability at this point in time would depend on the availability of four things: (1) adequate shelter, both from weather and from fallout; (2) minimally contaminated food and water; (3) adequate medical care for serious injuries; and (4) self-protection from further destructive forces, whether blast- or fire-induced or human in origin.

Dependent on time of year and location, climate is not likely to cause many overall deaths during this short period, though near the end of this immediate phase the lowered temperatures worldwide would begin taking a toll of the traditionally more susceptible, the young, ill, and elderly. Consuming contaminated food and water for several days or weeks would not cause a large number of immediate deaths, though longer range impacts could be very great. The inaccessibility of water and shortages of food would be felt within weeks, contributing to a general weakening of immediate survivors. Lack of adequate medical care would result in increasing numbers of deaths nearer the end of the immediate period, as supplies became exhausted and the multiplicity of injuries to victims combined to an irreversibly fatal conclusion.

There is little basis for speculation on the nature of individual response to such physical and psychological stresses as would cause irrational human behavior on a large scale, such as widespread attacks on other humans. It should be noted, however, that as time passes, competition for increasingly scarce resources essential to human survival would rise and with it the potential for asocial actions on an individual or concerted scale.

More profound implications for long-term societal survival arise from immediate-phase psychological responses to war and its aftermath. Some important separate psychological themes have been identified in victims surviving critical circumstances, especially relevant when related to survivors of the two atomic bombings of Japan in World War II.

Of initial critical importance to survivability of the individual is the state termed *psychic numbing*, or the limitation of the ability to feel (Lifton, 1967; Lifton et al., 1983). It can be described as a diminished capacity to feel, erected as a protective mechanism against full realization of overwhelming horror. A range of degrees of shutting off of unacceptable realities can enable an individual to respond in a world which would be otherwise personally unsurvivable. While originating in the immediate post-war period, this psychic state may last much longer, through varying levels of intensity and psychological importance.

Acting against the beneficial aspects of psychic numbing is the pervasive feeling of demoralization, reflected in apathy and an inability to act. Complicating actions taken particularly during this immediate time frame is the suffering by survivors of a prolonged sense of disorientation and a loss of connectiveness to other survivors (Erikson, 1976, cited in Chazov and Vartanian, 1982).

Of lesser importance in terms of physical survivability but of profound importance in shaping later societal systems are the individual's response to the indelible psychic imprint of images of death and the individual's capacity to function while being unable to erase these images. The feelings of survivor guilt seem universal, bringing psychological conflict to bear when recognizing great joy at being alive and concomitant horror at having been unable to save others.

These related disorders, directly caused by extreme crisis and wholesale destruction, have been identified with protracted anxiety states and reactive psychoses, which affect individual capacities to resist continued stress. It has been estimated that at least one-third of the surviving population would suffer from severe mental and behavioral disturbances (Chazov and Vartanian, 1982). How these psychological states would affect long-term survival rates is unknown, though obviously the reformation of societal functions is dependent on the will of people to survive and their psychological capacities for coping with long-term hardships.

Longer Term Post-War Effects. During the time period following the immediate effects of and responses to the nuclear detonations, needs of survivors and abilities to meet those needs would evolve. New social structures would also evolve to encompass those changed circumstances and activities.

Medical situations would both worsen and improve, in different respects. Those survivors who were severely injured by blast or were already critically ill would very likely have died, lessening demands on medical supplies, which would be, at this time, very scarce or nonexistent. In contrast, radiation sickness from fallout and from consumption of food and water contaminated by radiation would be placing increasing demands on long-term health care facilities and surviving medical personnel.

Malnutrition would increase as stored food sources become depleted and as the climatic conditions virtually preclude new production of foodstuffs. A diet largely consisting of grains and beans would induce chronic illnesses from protein and vitamin deficiency, treatment of which would be difficult or impossible. Inadequate diet and extremely cold temperatures would lower natural resistance to communicable diseases such as measles and diphtheria, and respiratory diseases such as streptococcal and pneumococcal infections and tuberculosis, for which immunizations or antibiotics would be scarce or nonexistent. Inadequate sanitation would lead to a host of enteric diseases, such as infectious hepatitis, cholera, amebic dysetery, and possibly typhoid. [For a thorough summary of the diseases to be experienced after a nuclear war, see Geiger (1981, 1983); Abrams and Von Kaenel (1981); Leaning (1982).] Uncontrolled insect growth, coupled with inadequate sanitation, would increase sharply the incidence of insect-vectored diseases, such as typhus and malaria, when ambient temperatures begin to increase. The diseases endemic to rural areas, where survivors would be migrating, would increase, and include rabies, plague, and tetanus; these would be exacerbated by population increases in those pest species (e.g., rats) that provide vectors for disease transmittance.

Urban and suburban migration to rural areas, to be closer to food sources and more remote from blast devastation, would cause conflicts between groups of survivors and those who never were within targeted areas and thus escaped the physical and psychological horrors of the blast-area residents. Disputes would inevitably arise over equitable distribution of remaining resources, particularly food and shelter. Blast survivors would have traveled from a world of immense horror, in rapidly decreasing temperatures and over a twilight landscape, in hopes of an increased chance of survival and an escape from seemingly inevitable and all-pervasive death and destruction. Rural residents, not having lived through such a series of ordeals, would be stronger physically and more likely better organized for mutual support, with a strong vested interest in maintaining and protecting assets. It would be highly speculative to project any potential resolution of these inevitable societal conflicts, except to say that more human casualties could be expected (see also, Lynch, 1983).

Economic systems would alter as dramatically as other elements of the overall societal shift. There would probably be only small differentiations in types of labor available, since basic survival through providing food and shelter would be central to every human endeavor. Monetary systems would be worthless without centralized government, and a barter and trade economy is the most likely evolution.

All human activities in this longer-term post-war period would be tempered by the ecosphere-level alterations of light and temperature. Subzero and subfreezing temperatures, coupled with extremely low light levels, would not only limit travel, communications, agricultural activities, and social interactions, but would, in a dark and icy world, change fundamentally life as it has been known historically, to a world beyond imagination. How human societies would adapt and survive speaks to the most basic level of existence of *Homo sapiens* on Earth.

Relatively Lesser Problem Areas

One of the amazing things about the effects of nuclear war is that virtually every environmental problem of concern to us today would be a direct part of nuclear war. As a partial listing, nuclear war would lead to habitat destruction, ozone (O_3) production, releases of oxides of nitrogen (NO_x), wildfires, emissions of toxic chemicals, release of radionuclides, siltation of surface water systems, acid precipitation, soil erosion, dam failures, species extinctions, and a host of other environmental problems. The only difference is that while we are now concerned about the cumulative and long-term effects from a myriad of anthropogenic stresses on the environment, the levels of those stresses from a nuclear war would in general far exceed current values, and they would occur almost instantaneously. Hence, each of those elements, if it were to occur absent all the other aspects of nuclear war, would constitute an environmental concern of virtually unprecedented importance; yet, considering the major problem areas discussed in previous sections, these problems pale in comparison. Nevertheless, they would have individually and collectively a dramatic effect on the quality of life for survivors, and, therefore, they must be addressed here. Further, many of the phenomena discussed in this section are the primary foci of previous studies of environmental consequences from nuclear war; thus, at a minimum, discussion is necessary for comparison across studies and to put the total consequences in proper perspective. In the remainder of this chapter, we will highlight many of these lesser issues.

$NO_x/O_3/UV$-B Effects. The issue of indirect effects on the ozonosphere has been of concern for the last ten years and was the major focus of the National Academy study on nuclear war (NAS, 1975). More recent analyses of NO_x production and alterations in the amount of ultraviolet (UV-B) light transmitted to the Earth's surface have supported the projections of the NAS study for a large-scale nuclear war in which warheads $\geq$ 1 MT are used (see Crutzen and Birks, 1982; Turco et al., 1983a; also, a new NAS study is currently underway).

Oxides of nitrogen are produced in the fireball by the extremely high temperatures followed by cooling to the point where reversal of the reactions does not occur. Crutzen and Birks (1982) and Crutzen (1983) provide a detailed discussion of the reactions involved in NO_x formation and subsequent reactivity. About 10^{32} molecules are generated per megaton of yield. A key element is the height of the mushroom cloud, as that determines how much of that quantity enters the stratosphere. In the stratosphere, NO_x acts to destroy ozone molecules. Projections are for about a 50% reduction in O_3 in the stratosphere under some nuclear war scenarios (Turco et al., 1983; Crutzen and Birks, 1982; NAS, 1975).

NO_x actually enhances O_3 production at low altitudes. Added to the O_3 produced directly by the fireball, local effects of excessive concentrations of O_3 in contact with plants can be significant (Table 29). Ironically, though, while O_3 incidents could be a problem, the depletion of O_3 in the stratosphere, resulting from interactions at high altitudes with the NO_x, would occur simultaneously.

Sunlight above the atmosphere contains intense ultraviolet light that would be very damaging to living organisms. Most of this light is absorbed in the upper atmosphere by the ozone layer, and only a fraction of a percent (0.1 to 0.5%) reaching the surface is in the UV-B waveband (280–320 nm). It is projected that this might increase by 50–400% after a large-scale nuclear war (see Table 29). This is somewhat deceptive, however, because the biological effect of light in this band is strongly dependent upon the wavelength. The effectiveness of a given radiation dose is a product of the intensity at each wavelength and its biological effectiveness. If the ozone layer were depleted so that the total UV-B radiation were to increase, most of the added radiation would be over a narrow range of wavelengths near 300 nm (Figure 26). These wavelengths are extremely damaging. The net result in terms of the biological effectiveness is a disproportionate increase in damage with only a small increase in the total UV-B dose. In the example shown (Figure 26), a 1% change in the total UV light would result in a 47% increase in DNA damage. Small changes in the amount of ozone present in the upper atmosphere, thus, can have disproportionately large effects on the biological consequences of UV-B radiation.

UV-B light is strongly absorbed by nucleic acids, aromatic amino acids, and the peptide bonds that are present in all organisms. Hence, UV-B has the potential to cause damage to all organisms. Some plants native to high elevations have evolved protective mechanisms (generally strongly absorbing substances in the epidermis) that would tend to protect them from damage. The damage would probably be most severe at lower elevations, especially at low latitudes which do not now receive much UV-B radiation. Photosynthesis of higher plants and algae is particularly sensitive to UV damage, and plants could receive heavy damage just as the aerosols cleared from the atmosphere, restoring the potential for photosynthetic productivity in natural and agricultural ecosystems. This could extend the interruption of primary productivity resulting from the period of smoke and dust.

Many forms of marine life are particularly sensitive to UV-B radiation and even small increases of UV-B have been shown to alter the nature of estuarine marine communities (Calkins, 1982).

Although the fractional increase of UV-B radiation would be less in the tropics than at temperate latitudes, the intensity of solar UV-B in the tropics is already much greater than at high latitudes (Caldwell et al.,

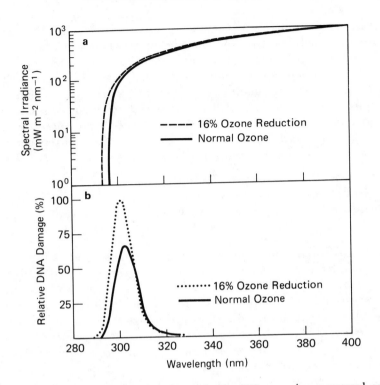

Figure 26. a. Global spectral irradiance in the UV assuming a normal ozone layer (*solid line*) or a 16% ozone reduction (*dashed line*). **b.** Relative DNA damage corresponding to the change in spectral irradiance. These damaged curves were obtained as the product of the action spectrum for DNA damage (Setlow, 1974) and the spectral irradiance curves of part A. (Derived from information in Caldwell, 1981)

1980). Thus, the increased UV-B in tropical ecosystems may be as severe a constraint as for temperate systems.

Other damage might occur:

1. The immune systems of *Homo sapiens* and other mammals are known to be suppressed even by relatively low doses of UV-B (Ehrlich et al., 1983). Especially under conditions of increased radiation and other physiological stresses, such suppression of the immune systems leads to an increase in the incidence of disease.
2. Protracted exposure to increased UV-B may lead to widespread blindness among humans and other terrestrial mammals (Pitts, 1982). In addition, the increased UV-B radiation might act synergistically with other stresses causing disproportionate effects. For

example, plant leaves that reach maturity under low light intensities (such as might precede the episode of increased UV-B) are two to three times more sensitive to UV-B than leaves that develop under high light intensities (Teramura et al., 1980; Warner and Caldwell, in press). Vegetation attempting to recover from low light and cold temperatures would be further constrained by enhanced solar UV-B radiation. Also, bacterial UV-B sensitivity is enhanced by low temperature, which suppresses the normal process of repair that is dependent on visible light (Ehrlich et al., 1983).

The strong direct effects of UV light and the likely synergisms that would occur with other stresses lead us to suggest that damage from increases in UV-B light may be among the most serious unanticipated consequences of nuclear war.

Hazardous Chemicals. Nuclear detonations and secondary fires would result in the release of large quantities of toxicants into the atmosphere and surface water systems. Examples include polynuclear aromatic hydrocarbons (PAH's) from forest fires; vinyl chlorides from structural fires; heavy metals from storage areas; biological agents from research and medical facilities; complex hydrocarbons from fuel storage areas; a variety of toxicants from fires at industrial facilities; massive spills from liquid and gaseous chemical storage tanks; and many other sources of toxicants. There is much too great an uncertainty to allow adequate quantification of the amounts of such toxic chemicals that would be released to the environment. However, experiences in the last few decades in the U.S. and other developed countries have shown that relatively small releases from accidental events (e.g., kepone in Virginia, PCB's in Italy) have resulted in large tolls on humans and the environment. It could be expected that in localized areas, highly toxic chemicals could render areas uninhabitable by humans and other species. However, in the aftermath of a nuclear war, survivors would not likely detect and avoid many of these areas until adverse health effects became apparent, which would not happen readily in a world with so many other human and environmental stresses.

Air Pollution. In addition to the $NO_x/O_3/UV-B$ issue, a number of conventional air pollutants would be released in large quantities. Included are CO_2, CO, acid precipitation precursors (e.g., $SO_3^=$), and constituents of smog. The near-surface O_3 levels, for instance, could be high enough (Table 29) to result in widespread damage to forest and crop species, with average values for the Northern Hemisphere remaining up to 150 ppbv for three months after a nuclear war (Turco et al.,

1983a, b). This average value is only a factor of two or three below the two-hour exposure levels that are thresholds for significant plant injury (Treshow and Stewart, 1973). Also, these levels perhaps could cause physiological effects for sensitive humans respiring O_3-laden air. Similarly, NO_2 is produced by nuclear detonations in substantial quantities. Exposure to only 150 ppm of NO_2 for very short periods of time can cause human fatalities (Health and Welfare Canada, 1981; NAS, 1977), and can adversely affect many species of deciduous and coniferous trees (Smith, 1981). For both O_3 and NO_2 exposures, many plants demonstrate significant growth reductions in the absence of apparent foliar injury (Kress and Skelly, 1982).

The CO_2 releases would not lead to direct health effects on humans or other species; however, a somewhat closer look is required for the potential for long-term climatic alterations from changes in CO_2 concentrations in the atmosphere.

Direct releases of CO_2 from forest fires could be on the order of $1-2.5 \times 10^9$ tons C (Crutzen and Birks, 1982), or about 1% of the temperate forest biomass of the Earth (Bolin, 1977). CO_2 emitted to the atmosphere will not remain in the atmosphere, with substantial fractions being absorbed into abiotic cycles in the oceans and into photosynthetic fixation. Under current conditions, less than half of the CO_2 emitted to the atmosphere from fossil fuels remains to increase the atmospheric concentration, based on observed airborne fraction of fossil fuel carbon of 0.505 at the South Pole and 0.548 at Mauna Loa, HI (Bacastow and Keeling, 1981).

However, reduction of rates of carbon fixation into forest biomass would result from reduction of photosynthetic activity from light and temperature decreases. Estimates of current carbon fixation rates for temperate forests range from $1-1.2 \times 10^9$ ton C yr^{-1} (Armentano and Ralston, 1980) to 1.1×10^{10} ton C yr^{-1} (Whittaker and Likens, 1973; Bolin, 1977). (The lesser number is probably more accurate, as it considers actual forest inventory data rather than potential vegetation.) Thus, if photosynthesis were essentially curtailed for the first year after a nuclear war, 10^9-10^{10} ton C would not be fixed by forests; additional figures would relate to reduced productivity in agricultural and oceanic ecosystems.

However, in actual practice, agricultural and commercial forestry activities are responsible for net releases of carbon through depletion of soil carbon and litter following harvest (Moore et al., 1981; Cropper and Ewel, 1983). Further, in the aftermath of a nuclear war, fossil fuel burning would be greatly reduced as, for example, industrial and domestic consumption of coal and oil supplies could be expected essentially to cease. Estimates of current fossil fuel releases of CO_2 are about 5×10^9 ton C yr^{-1} (Emmanuel et al., 1980; Rotty, 1979), perhaps five times the rate of forest fixation. Finally, decomposition of litter, which returns substantial amounts of carbon to the atmosphere, would likely slow considerably during the period of extreme cold temperatures; but once the temperatures had returned closer to normal values, decomposition rates could exceed current values as all the biomass from recently killed plants and animals decay.

The net result of these opposing forces could lead to either increased or decreased CO_2 concentrations in the atmosphere. Uncertainties here are too large, since there are large-scale potential effects in each direction. There are no compelling reasons to believe CO_2 levels would be changed to the point of altering the thermal radiative balance of the atmosphere after a large-scale nuclear war.

Habitat Loss. An important ecological issue is the effect from the direct destruction of ecological habitats by detonations, secondary fires, floodings, etc. As discussed previously, blast effects would not cause widespread habitat destruction, except perhaps for estuarine and coastal areas. Flooding from dam ruptures could be quite extensive, and wildfires could cover a substantial fraction of ecosystems. Indirect effects could be expected to be important, particularly involving human stresses; for example, increased demand for firewood could result in signficant pressures on forests and desert systems.

The ecological consequences of nuclear war-induced wildland fires are impossible to predict from our current knowledge of fire ecology, largely because the climatic conditions and spatial extent of affected areas would be unprecedented in the aftermath of nuclear war. Under current conditions, prescribed fire is gaining recognition as a valuable management tool for forests as well as grasslands and shrub-steppe vegetation types (Kozlowski and Ahlgren, 1974; Wright and Bailey, 1982). In general, increased vegetative growth following fire results in higher densities of large and small mammals in the affected area (Ream, 1981; Wright and Bailey, 1982). Reduced solar insolation and severely lowered atmospheric temperatures would limit vegetation recovery, thereby suppressing recovery of higher tropic levels. Darkened skies would limit the soil warming, often observed after wildland fires, that is important to increased nutrient mineralization by microorganisms and also to plant seedling establishment. In the absence of vegetative recovery, exposed soils would be more susceptible to erosive processes, and additional nutrients could be lost from these areas as a result (Woodmansee and Wallach, 1981). Hence, for those areas subjected to nuclear war-induced wildland fires, the long-term consequences would be more severe than are generally assumed under current conditions because recovery would be much slower and therefore limited.

Extinctions. Clearly the potential climatic alterations suggested here are capable of causing widespread extinctions of species, particularly in terrestrial and freshwater systems. Tropical ecosystems might be particularly vulnerable to extinctions from temperature reductions (Ehrlich et al., 1983). Analogies to previous extinction events are possible.

The extinction of particular species can have significant effects through a number of mechanisms:

1. Loss of *important species*. This situation is where a species that has particular direct importance to humans is lost. Examples of important roles are for human food, for development of medicines, and for energy or shelter resources. The classification of a species as having direct importance for humans, of course, would alter in the aftermath of a nuclear war, as humans sought replacement resources for necessities currently provided by industrial and agricultural systems. Extinction of these species could directly reduce potential human survival.
2. Loss of *critical species*. These are the species that play some particular role in the maintenance of the overall biological community, so that loss of such a species would lead to effects on other species that are dependent on it. An example of a critical species loss is the extinction of predators, such as birds, that control more rapidly recovering insect species. Thus, while the bird species per ser may not provide direct support to humans, its loss could lead to outbreaks of pests that are of serious concern to surviving humans. Another example of a critical species is the loss of key insect pollinators needed for plant reproduction.
3. Loss of *functional species*. Certain species or group of species, particularly microorganisms, may control ecosystem-level processes that are essential for natural system survival and recovery and, thus, indirectly necessary for human survival. Extinction of species that have such a functional role could lead to widespread effects on ecosystems and, thus, on human survivors.

Genetic Diversity. For species surviving the effects of nuclear war, their ranges may be severely reduced, and there may be differential effects on particular ecotypes or genotypes. This phenomenon could have major implications for human and natural system recovery.

Genetic diversity in every species helps provide flexibility for that species to adapt to different environmental conditions. The imposition of severe stress, particularly that to which the species had not been adapted, could allow only a small subset of the potential genetic types to survive, thereby reducing the genetic diversity within that species. Similarly, loss of whole species results in loss of genetic diversity across species that perform similar functional roles. The key issue in either type of genetic diversity decreases is whether or not sufficient flexibility remains for ecosystem functioning to continue and recover, given the continual environmental perturbations following a nuclear war for many months or years.

Species Diversity. One of the classical responses by communities to extreme stress is simplification, and reduced diversity of constituent species. Insofar as this reflects a reduction in the functional redundancy of ecosystems, this may significantly affect the functioning and stability of natural ecosystems as well as rates of recovery. Additionally, this issue is germane to outbreaks of opportunistic, pest species from loss of predatory control, and consequential impacts on human health and food production. At a minimum, the combination of reduced genetic diversity and widespread species extinctions can be expected to result in long-term reverberations throughout the biological systems, as new species interactions become established in the aftermath of a nuclear war. How the resultant biological systems would perform with respect to stability and with respect to human support is highly uncertain.

Nutrient Dumping. Extensive fires, erosion, forest felling, flooding, and other events likely to result from nuclear war can cause the export of large portions of the available nutrient pools in terrestrial systems by surface and groundwater runoff. This could have potential impacts on the successional development of terrestrial ecosytems during recovery periods.

Groundwater Contamination. Various radioisotopes, hazardous chemicals, biological agents, and other toxicants released as a result of nuclear war can enter unconfined aquifers, contaminating underground sources of drinking water (Naidu, 1984; Wetzel, 1982). Contamination would be worse in near-surface aquifers, where increased human utilization would occur because of lack of energy to extract water from deeper sources. While the quick freezing of water near the surface would limit aquifer recharge in the short run, when the frozen surface thawed, some of the soluble radionuclides could enter groundwater systems. We could expect to see quantities of ^{90}Sr, ^{129}I, and ^{137}Cs in groundwater for decades after the war. If societies recover sufficiently to be able to exploit groundwater from shallow to moderate depths, they could continue to be subject to low levels of radionuclides and chemical toxicants.

Bioconcentration of Radionuclides. In previous sections, we have discussed the effects on humans of long-term fallout, including internal doses received by consumption of food and water contaminated with radionuclides. An extensive literature exists on the specific phenomenon of bioconcentration by biota of radionuclides and some toxic chemicals. Such occurrences could lead to substantially increased exposures to internal radiation and chemical toxicity for humans consuming food from certain bioconcentration chains.

One of the important aspects of radionuclide bioconcentration and incorporation into human food webs is that the longer lived radio-nuclides (i.e., the ones that exist long enough to reach humans) are often alpha-emitting radionuclides. As discussed previously, the relative biological effectiveness of alpha particles is 10–20, so that the same absorbed dose of alpha particles as of gamma rays is 10–20 times more deleterious. For exposure to fallout, alpha-emitters are not important since the alpha particles are absorbed in very thin layers of tissue, thus not penetrating the skin. However, ingestion of alpha-emitters via consumption of contaminated food allows the alpha particles direct access to important human tissues, such as in the digestive, circulatory, and bone systems.

A similar situation exists for the less damaging but more prevalent beta-emitters, such as cesium and strontium, which are intermediate in duration but highly active biologically. Thus, the ingestion of beta- and alpha-emitting radionuclides is very important to long-term human health.

Uptake of these radionuclides from air and water by biota brings them into the internal structure of plants and animals. Hence, washing the surface of foodstuff does not eliminate the source of radiocontamination for human consumers. Further, the process of bioconcentration (i.e., active biological processes that cause the concentrations of the radio-nuclides in tissues to exceed the concentration in the source, such as water) significantly enhances the danger to humans.

Bioconcentration of toxic chemicals poses a similar problem, though on a much more localized scale.

Other Climatic Effects. In addition to the dramatic alterations in light levels and surface temperatures, a variety of ancillary climatic effects could ensue. Included are changes in albedo from deposition of particulates on polar regions; effects of deforestation on albedo and on evapotranspiration; storms at the continental/marine interface; alterations in patterns of ocean currents and associated effects on nutrient movements via changes in upwhelling; changes in the hydrologic cycle; local effects from average global changes; floods, avalanches, and a number of other phenomena.

Differential Vulnerability of Species. Some generalities can be detailed concerning the relative vulnerability of different types of species. For instance, for plants, it is critical where the perennating tissues are located: belowground results in considerable protection from exposure to cold, wind, radiation, etc., whereas terminal buds are particularly exposed. The adaptive strategies of species is important: e.g., opportunistic species are more capable of surviving extreme conditions, and they are adapted to rapid population and biomass increases as competition from more sensitive species is curtailed. Most species have differential vulnerabilities, and often differential exposures to stress, among the various stages during life cycles; often juvenile and larval stages are particularly sensitive, but sometimes this can be compensated for on the population level. Another aspect is that marine species in general would be significantly protected from the extreme environmental conditions, as the water provides considerable buffering from changes in temperature and light levels, greatly decreases the distances for which radionuclides can cause external doses, and provides isotopic dilution. On the other hand, coastal systems may have the longest exposure periods to long-lived radionuclides because of erosion, differential sorption to silt and other particulates that eventually reach coastal areas, and other factors. Concurrently, there is a downstream magnification of pollutant loadings. In summary, a variety of characteristics of biota would have much to do with the species-level responses to extreme conditions, and, consequently, with the nature of surviving biota and their utility for surviving humans.

Pest Outbreaks. An issue related to the differential effects on opportunistic vs. sensitive species is the potential for significant population explosions for species that are pests to humans, particularly insects and rodents. As discussed previously, it can be anticipated that the fluctuations in the biological community structure that are almost certain to occur after a nuclear war and nuclear winter will likely include frequent, large-scale outbreaks of opportunistic pest species for which humans would have little capability to control. Effects on humans relate to competition for food sources (e.g., explosion of herbivorous insects); transmission of diseases (e.g., rats and associated fleas carrying plague); and many other factors.

5

Recovery Processes

In the consequence analyses thus far, we have addressed the human and environmental effects, including the physical and biotic responses to a nuclear war of a scale large enough to cause widespread effects. The atmospheric responses are projected to result in peak reductions in temperature and light after a few weeks. Climatic changes would gradually return to pre-nuclear war conditions over a period of several years. There is no compelling reason to believe the atmosphere would not essentially recover after a decade or so post-war.

During that period, however, tremendous changes in human and ecological systems would occur. Foremost among these, approximately one-fourth of the world's population could be direct casualties of the nuclear war in the early time period, and longer term casualties from disease, starvation, and societal disruptions could number in the billions. At some time after the climatic effects peaked, human life losses would peak, and the population of *Homo sapiens* would reach some minimum value; exactly how low that value would be cannot be determined. For now, we assume there are humans surviving beyond some time at which losses reach a maximum value. In this post-maximum casualty period, the nature of human existence would be inextricably tied to the processes controlling human and ecological recovery.

A primary consideration here is the strong relationship that would exist between human and ecological systems (Harwell, 1984). As we have discussed, a nuclear war of large scale would lead to the essential termination of human support systems for much, if not all, of the world. The societal systems that currently allow the Earth to support almost 5×10^9 people would not be able to survive the loss of billions of people, the disruption of transportation, energy, and communication systems, the ravages of climatic changes on agricultural systems, as well as a host of other situations.

Consequently, those humans who survive the direct effects would be forced to rely on natural systems for their sources of food, shelter, energy, water, and other life essentials. Yet those natural systems themselves would have experienced widespread, intense insults, affecting the survival of large numbers of species and basic ecological functions such as productivity. As we have seen, these insults could lead to near-zero production in agroecosystems for a year, with substantial reductions resulting from climatic effects for some time thereafter. Similarly, primary productivity in natural terrestrial and freshwater ecosystems could also be reduced to very low levels for many months. How drastic these ecological responses cumulatively became would have much to say about the eventual surviving human population levels.

In the post-maximum casualty period, recovery rates of human population levels would be affected by the rates of recovery of productivity of natural systems. This might take longer than climatic recovery because of time lags resulting from successional processes that were slowed because of nutrient losses, soil erosion, reduced species and genetic diversity, habitat destruction, and lingering toxic chemicals in the environment. As the FORNUT simulations showed, even absent these factors, decades could pass before deciduous forests recovered productivity after experiencing just a few years of reduced temperatures. Other terrestrial systems could have similar responses, although grasslands could be expected to recover productivity essentially as fast as climatic conditions recovered. Likewise, freshwater ecosystems could in general recover more rapidly than forests, being limited by climatic conditions and secondarily by the state of the watershed ecosystems that provide their inputs. Coastal and estuarine ecosystems could recover rapidly with respect to water column conditions, because of the rapid response times of plankton, but if extensive damage to bottom habitats occurred, recovery could be prolonged; further, these systems are the last in the long chain of systems through which hazardous chemicals, particulates, radionuclides, etc. pass.

Atmospheric systems, then, would recover first, followed after time lags by natural biological systems. Human systems could only at best track the rates of recovery of the latter. This is because the minimum level of human population reached could be largely determined by the carrying capacity of the biosphere in its stressed condition. That capacity would first have to recover before human populations could.

But other factors would operate on the recovery rates for *Homo sapiens*, specifically the redevelopment of human support systems. Important here is the reestablishment of agricultural systems; these are at the juncture of biological and human recovery processes since they fundamentally require both. Factors include sources and availability of seed, fertilizers, transportation, noncontaminated land, and energy supplies. Other important factors would be the redevelopment of economic systems, as barter becomes established and eventually replaced by money systems; the redevelopment of energy, transportation, and communication systems; and the reestablishment of medical systems. In each of these, the differentiation of human labor is required, quite in contrast to the initial responses to nuclear war, in which most or all individuals would likely have to spend all their energies strictly on the survival functions, such as acquiring food, warmth, and shelter. Here is a situation where there is a strong feedback from the state of ecological systems to the state of human systems: Only after natural systems provide some excess of support could there be the freedom for individuals to develop the specialized activities necessary for division of labor and evolution of societal systems.

There would be feedback in the other direction, too. The exploitation of natural systems by humans searching for food, water, and fuel would be expected to retard the recovery of the natural systems. The nature of these stresses would change as societal support functions evolved, completing the feedback cycle.

Other factors affecting human recovery relate to psychological damage and the resultant life perspective of survivors. Perhaps new religions or other cultural responses could ensue that could have an effect on societal recovery. The degree of loss of technology during the period of recovery would be important. Obviously all information and knowledge would not be lost, but much technological support would be no longer available.

The rates of coalescing of humans into organized groups could be affected by opposing forces: In one sense there is security and sustenance for an individual by joining with others in a common effort for survival; opposing this could be the disruptive forces of competition for resources among unconnected groups. As an example of the implications of this, there may be conflicts between nomadic peoples, whose survival strategies are based on exploitation, mobility, and renewed exploitation across the landscape, and those people whose strategy is to establish at a single location the societal and sustenance systems necessary for continual survival. The latter's strategies would include storage and protection of resources; the former's could include taking from the labors of others, in a replay of some of the major cultural conflicts of past civilizations (Bronowski, 1978).

Predicting the nature of societal systems is highly uncertain. Our intent here has been just to identify some of the major factors that could contribute to or retard societal recovery. The very existence of this large uncertainty, however, is important in one sense: Conducting a nuclear war to effect some particular result that is desired by a warring country would lead to unpredictable, uncontrollable results that are not likely to be those desired.

How far human and ecological systems would regress after a nuclear war, how rapidly the biological and atmospheric systems could return to near normal conditions, how the mix of factors working towards or against social reorganization and reestablishment of human support systems would interplay, and how the synergisms and feedbacks among all of these forces would operate, all would be major factors in defining the quality of human life in the long term after a nuclear war. It could be expected that no quick recovery is possible, that many forces would act to retard recovery, and that the human effects of a large-scale nuclear war would persist for long periods of time.

6

Summary of Consequences

In the preceding sections of this book, we have presented a wide array of individual effects that could result from a large-scale nuclear war. We have included direct, immediate consequences, particularly from blast, thermal radiation, and initial ionizing radiation. Immediate indirect effects, such as from secondary fires, were discussed, as were longer term effects, especially from temperature reductions, light reductions, fallout, societal disruptions, and a suite of other environmentally mediated mechanisms. The purpose of the present chapter is to provide a brief picture of all of these factors in order to convey a comprehensive image of the world after nuclear war.

It should be recalled that the analyses performed here are based on considering a large-scale nuclear war, defined as a war with sufficient detonations to result in the types of atmospheric effects of temperature and light reductions suggested by Crutzen and Birks (1982) and projected in detail in Turco et al. (1983a,b). The particular scenarios analyzed were selected and quantified as representing the suite of scenarios that could be envisioned for counterforce and countervalue nuclear war strategies. Scenarios of greater or lesser consequences could be developed.

For the scenario analyzed here, and its parametric alterations, the immediate and longer-term consequences have been represented in Figure 27 and Tables 37 and 38. The immediate effects result primarily from blast, thermal radiation, initial ionizing radiation, and fires resulting from the detonations themselves. These phenomena would occur during the immediate minutes to hours of the nuclear exchange. The resultant human casualties would occur in the early stages of the post-war world. A summary of the magnitude of the consequences is provided in Table 28 (page 83), where it is reported that 50–90 million deaths and 30 million injuries could occur in the United States alone, primarily including those individuals within the lethal area of blast. The

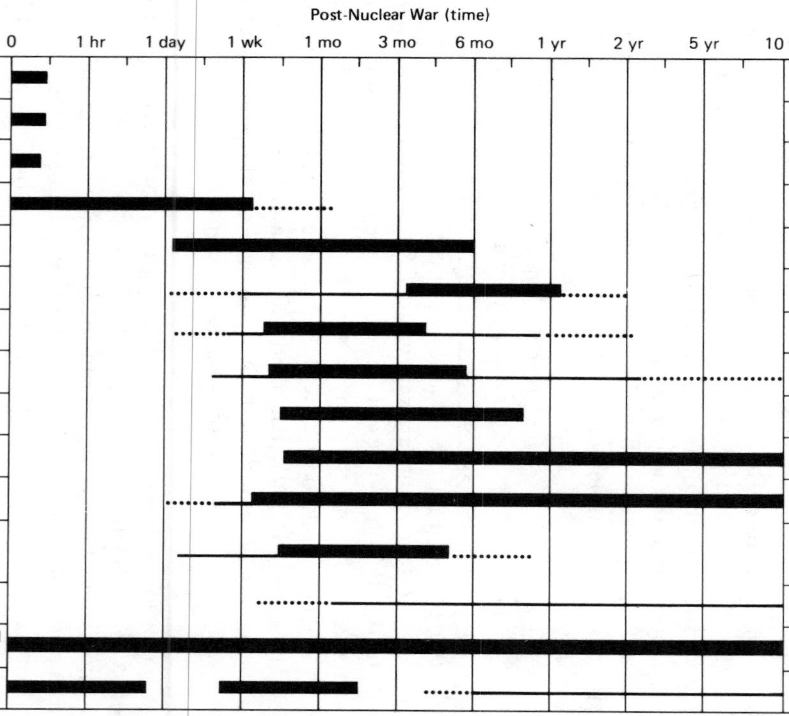

Mechanism of Effect	Post-Nuclear War (time)	Population at Risk			Casualty Rate for Those at Risk	Potential Global Deaths[4]
		U.S.[1]	N.H.[2]	S.H.[3]		
Blast		H	M	L	H	M-H
Thermal Radiation		M	M	L	M	M-H
1° Ionizing Radiation		L	L	L	H	L-M
Fires		M	M	L	M	M
Air Pollution		H	M	L	L	L
Stratospheric O$_3$ Reduction		H	H	M	L	L
Light Reductions		H	H	M	L	L
Temperature Reductions		H	H	H	H	M-H
Frozen Water Supplies		H	H	M	M	M
Food Shortages		H	H	H	H	H
Medical System Collapse		H	H	M	M	M
Diseases: Contagious (shelter period)		M	M	L	H	M
Epidemics (pest vectors)		H	H	M	M	M
Psychological/Societal Stress		H	H	L	L	L-M
Radiation		H	H	L-M	M	M

[1] U., S., United States [2] N. H., Northern Hemisphere [3] S. H., Southern Hemisphere [4] Global deaths: L, 0-10^6; M, 10^6 -10^8; H, 10^9

Figure 27. Summary of timing and magnitude of direct and indirect effects from a large-scale nuclear war.

total casualties here, of 80–110 million people, are for a targeted population of about 120 million people (i.e., all of those in the top 300 cities), suggesting that a very large fraction of the U.S. urban population would be dead or seriously injured essentially instantaneously.

This same time period would see the essential termination of organized societal functions in the targeted urban areas of the U.S. Concomitantly, the loss of communications and energy systems would likely occur from the effects of the electromagnetic pulse (EMP) (Lundquist, 1983) and from direct blast effects, so that from the early moments of a large-scale nuclear exchange, little accurate information would likely be available to survivors.

Among the urban populations, there would be large numbers of injured who would be subject to flash blindness, shock, and collapsed structures, all precluding ready escape. Related to this, many would be separated from their families and friends with great difficulty in finding them or even knowing of their fate. The psychological imprinting of unprecedented death would begin during this period.

Within the initial few hours, secondary fires would grow in extent and intensity, resulting in deaths of those within firestorm areas who could not escape because of being injured or entrapped. These fires would force mobile individuals to escape outward from devastated areas, initiating a long-term trend in radiating outward to less impacted areas. This could enhance separation of individuals and fractionation of groups of people. These same fires would provide the seeds of a new devastation to be experienced in the weeks and months later.

After some period of time had elapsed, the medical system of the U.S. would be saturated and then overwhelmed with blast and burn casualties orders of magnitude greater than the capacities of the surviving medical facilities and personnel. Services would quickly degenerate into just distribution of pain killers, water, and minimal contact with medical staff. Injuries that today could be treated successfully would typically result in death.

Within minutes to a few days, local fallout would begin to be deposited, eventually covering one-quarter or more of the continental U.S. with lethal levels of radiation. Even including a protection factor typical of indoors habitation in nonurban areas, fully 12–18 million people could receive in the first two days a dose sufficiently large to assure their death within the next one or two months. Unlike those who are doomed to die of burn injuries, these people would not know of their fate. Only after some days would they experience the anxiety and nausea symptoms of early stages of lethal acute radiation, symptoms likely for many or most survivors of such devastation. Further, an additional 40 or

Table 37. Potential Ecological and Human Impacts from Climatic Changes Induced by Nuclear War.[a]

	Time after nuclear war	
First few months	End of first year	Next decade

I. *Natural Ecosystems: Terrestrial*

First few months	End of first year	Next decade
Extreme cold, independent of season and widespread over Earth, would severely damage plants, particularly in mid-latitudes in the Northern Hemisphere and in the tropics. Particulates obscuring sunlight would severely curtail photosynthesis, essentially eliminating plant productivity. Extreme cold, unavailability of fresh water, and near darkness would severely stress most animals, with widespread mortality. Storm events of unprecedented intensity would devastate ecosystems, especially at margins of continents.	Many hardy perennial plants and most seeds of temperate plants would survive, but plant productivity would continue to be depressed significantly. As the atmosphere clears, increased UV-B would damage plants and impair vision systems of many animal species. Limited primary productivity would cause intense competition for resources among animals. Many tropical species would continue to suffer fatalities or reduced productivity from temperature stress. Widespread extinction of vertebrates.	The basic potential for primary and secondary productivity would gradually recover; however, extensive irreversible damage to ecosystems would have occurred. Ecosystems structure and processes would continue to respond unstably to perturbations, and long period of time might follow before functional redundancies would reestablish ecosystem homeostasis. Massive loss of species, especially in tropical areas, would lead to reduced genetic and species diversity.

I. Natural Ecosystems: Aquatic

Temperature extremes would result in widespread ice formation on most freshwater bodies throughout the Earth, particularly in the Northern Hemisphere and in mid-latitude continental areas. Marine ecosystems would be largely buffered from extreme temperatures, with effects limited to coastal and shallow tropical areas. Light reductions would essentially terminate phytoplankton productivity, eliminating the support base for many marine and freshwater animal species. Storms at continental margins would stress shallow-water ecosystems and add to sediment loadings. Potential food sources would not be accessible to humans or would be contaminated by radionuclides and toxic substances.

Early loss of phytoplankton would continue to be felt in population collapses in many herbivore and carnivore species in marine ecosystems; benthic communities would not be as disrupted. Freshwater ecosystems would begin to thaw, but many species would have been lost. Organisms in temperate marine and freshwater systems adapted to seasonal temperature fluctuations would recover more quickly and extensively than in tropical regions.

Recovery would proceed more rapidly than for terrestrial ecosystems. Species extinctions would be more likely in tropical areas. Coastal marine ecosystems would begin to contain harvestable food sources, although contamination could continue.

[a]Table from Ehrlich et al. (1983); copyright 1983, American Association for the Advancement of Science.

Table 37 (*cont.*)

	Time after nuclear war	
First few months	End of first year	Next decade
	II. *Agroecosystems*	
Extreme temperatures and low light levels would preclude virtually any net productivity in crops anywhere on Earth. Supplies of food in targeted areas would be destroyed, contaminated, remote, or quickly depleted. Non-targeted importing countries would lose subsidies from N. America and other food exporters.	Potential crop productivity would remain low because of continued though much less extreme temperature depressions. Sunlight would not be limiting but would be enriched with UV-B. Reduced precipitation and loss of soil from storm events would reduce potential productivity. It is unlikely that organized agriculture would occur, and modern subsidies of energy, fertilizers, pesticides, etc., would not be available. Stored food would be essentially depleted, and potential draught animals would have suffered extensive fatalities and consumption by humans.	The biotic potential for crop production would gradually be restored. Limiting factors for reestablishment of agriculture would relate to human support for water, energy, pest and disease protection, etc.

III. *Human/Societal Systems*

Survivors of immediate effects (from blast, fire, and initial ionizing radiation) would include perhaps 50–75% of Earth's population. Extreme temperatures, near darkness, violent storms, loss of shelter and fuel supplies would result in widespread fatalities from exposure, lack of drinking water, and multitudes of synergisms with other impacts, such as radiation exposure, malnutrition, lack of medical systems, psychological stress, etc. Societal support systems for food, energy, transportation, medical care, communications, etc., would cease to function.

Climatic impacts would be considerably reduced, but exposure would remain a stress on humans. Loss of agricultural support would dominate adverse human health impacts. Societal systems could not be expected to be functioning and supporting humans. With the return of sunlight and UV-B, widespread eye damage could occur. Psychological stresses, radiation exposures, and many synergistic stresses would continue to impact humans adversely. Epidemics and pandemics would be likely.

Climatic stresses would not be the primary limiting factors for human recovery. Rates of reestablishment of societal order and human support systems would limit rates of human population growth. Human carrying capacities could remain severely depressed from pre-war conditions for a very long period of time, at best.

Table 38. Potential Ecological Consequences of Nuclear War Other Than Induced by Temperature and Light Reductions.[a]

Stress	Intensity/Extent	Mechanisms of effects	Ecosystem consequences
Local, global radioactive fallout from nuclear detonation and nuclear facilities	≥100 rem avg. background; ≥200 rem over large area in N. H.	Direct health effects; immune system depression; differential radiosensitivities of species; genetic effects	Alteration in trophic structures; pest outbreaks; replacement by opportunistic species; genetic and ontogenetic anomalies
Enhanced UV-B	4-fold increase over N.H.	Suppression of photosynthesis; direct health effects; differential sensitivities of species; damage to vision systems; immune system depression	Reduction in primary productivity; alterations in marine trophic structures; blindness in terrestrial animals; behavioral effects of insects including essential pollinators
Fire	Secondary fires widespread over N. H.; ≥5% of terrestrial ecosystems affected	Direct loss of plants; damage to seed stores; changes in albedo; habitat destruction	Deforestation and desertification which continues via positive feedback; local climatic changes; large-scale erosion and siltation; species extinction; nutrient dumping
Chemical pollution of surface waters	Pyrotoxins; release from chemical storage areas	Direct health effects; differential sensitivities among species; bioconcentration	Loss of organisms; continued contamination of surface and groundwater systems; loss of water for human consumption
Chemical pollution of atmosphere	Major releases of NO_x, O_3, and pyrogenic pollutants from detonations; major releases of toxic organics from secondary fires in urban areas, chemical storage facilities	Direct health effects; differential sensitivities among species; acid precipitation	Widespread smog; freshwater acidification; nutrient dumping

[a]Table from Ehrlich et al. (1983); copyright 1983, American Association for the Advancement of Science.

50 million Americans would experience over the next few weeks enough of a dose from exposure to local fallout that their deaths would be assured over subsequent months. Again, they would likely not know of their exposure until it became too late, as few of these millions would have the means or understanding to determine potential dose burdens, identify contaminated food, water, and land, and so on. What remained of medical support after burn and blast injuries had succumbed would be overwhelmed by terminal radiation-ill patients for whom few mitigating measures could be performed.

The end result of these immediate effects could be one-half to three-fourths of the U.S. population as eventual fatalities, and another 15–20% as injuries. On a worldwide basis, fully one billion deaths and one billion injuries could ensue. The population of the Northern Hemisphere would, of course, suffer virtually all of these casualties.

The fires started by the war would continue to burn in many urban and natural areas for many days after the exchange. Smoke and soot would locally stress humans and other biota. However, several more days would lead to a blanketing of most of the Northern Hemisphere with dense clouds of smoke and soot, reducing noontime light levels to near darkness. This would begin the series of ecological effects that would persist for years.

First would be a substantial reduction in photosynthetic activity by crop plants, forests, and marine phytoplankton, down to levels below the compensation point. Crop productivity from this alone could be substantially reduced, and large-scale phytoplankton population crashes could ensue. Forest trees could weather this lack of light by relying on stored energy reserves.

However, another result of reducing light levels would be major changes in the air temperatures over all of the Northern Hemisphere, especially in mid-continental areas. Temperature extremes would peak after several weeks at levels so low as to cause widespread death of trees and crop plants, wildlife, and, of course, unprotected humans.

But by the time surviving humans would have to face such low temperatures, large-scale migrations from their pre-war residences would have been required. This follows from the depletion of food reserves in non-farm situations within ·a few weeks and the forced emigration to areas where food, especially grains, are stored. Associated with such migrations would be exposure to local fallout at periods of substantial doses, difficulties in transportation, lack of good information on where to go to find food, and probable conflicts as those fleeing targeted areas compete for food stored by those who did not experience any nearby detonations. During this period, it could be expected that effective governmental control and social order would be lost in targeted countries. For those countries not involved in the nuclear exchange, within a few weeks the cessation of food imports, especially from North America, would likely result in similar food shortages, fleeing of cities, and competition for food reserves. This would be the precursor to the death of millions, perhaps billions, in targeted and nontargeted countries alike as food resources available to survivors became depleted and as no new agricultural production occurred.

Thus, the period of peak darkness and peak cold would coincide with the beginnings of starvation. Competition for food would be repeated in competition for fuel, clothing, and shelter in protection from the cold. Further, such sudden, drastic temperature changes could spawn extreme weather, with storms of unprecedented scale and intensity possible. It is during this period that many radiation-induced deaths would occur.

Within a few days to weeks after the real cold began, freshwater supplies would freeze over. In fact, at its peak, virtually all freshwater systems would freeze to 1 m or more thickness of ice. Water could become limiting for people and animals.

During these early days between the war and the end of extreme cold periods, people would have been closely grouped in shelters for protection from radiation, cold, and nomadic groups. Concurrently, sanitary conditions would deteriorate as uncontaminated water became rare and human wastes built up. Consequently, infectious diseases, such as dysentery, influenza, and cholera, would become widespread. Essentially little medical support would remain, and fatalities from such diseases could include a substantial fraction of survivors. The extremes of psychological stress, physical exposure, injuries, inadequate diet, sublethal doses of radiation, and other such factors would exacerbate susceptibility to disease. Here, as in the case for radiation-induced casualties, the young and the elderly would suffer disproportionately.

There would be other environmental stresses on humans. Air pollution, especially ozone, could reach levels sufficient for health effects of sensitive individuals. Toxic chemicals released by blasts or subsequent fires could cause localized stresses. Somewhat later, as smoke and soot began to clear from the air, the renewal of sunlight would bring the unwanted light at UV-B wavelengths, perhaps increasing many-fold at the most adverse frequencies. Humans could suffer blindness, and the initiation of melanomas could occur. Other species could die from the ultraviolet radiation.

And as the sun came back out and people emerged from seclusion, diseases of another sort would become widespread, especially those borne by insect and rodent vectors. Epidemics could result from diseases such as the plague, and many millions more could succumb. The incidence of disease in particular groups of people could enhance the friction with other groups, as, for example, nomadic groups in search of food and energy but carrying disease contact more sedentary populations. Social strife during these periods, which could persist for long periods of time, could remain high.

Over the longer term, as temperatures and light levels recovered to near pre-war conditions, humans would be forced to rely on natural systems for sustenance as their food reserves became further depleted. But the first year of the post-war world would see essentially no capability for agricultural productivity, and only a partial capability (with respect to environmental conditions) in the subsequent few years. Likewise, the productivity of terrestrial ecosystems would be essentially curtailed in the first several months, and some time lag would occur before substantial support from natural terrestrial ecosystems for food could exist.

Freshwater ecosystems might take another year to thaw, and they would largely be reduced in animal populations. Human predation may well more than keep pace with the recovery of freshwater biota, thereby retarding that recovery substantially. During this period, little reliance could be placed on coastal and estuarine ecosystems because of habitat destruction, disproportionate loadings of radionuclides and other toxicants, and likely extreme weather conditions at this boundary between continental and maritime areas, with concomitant high lateral gradients in air temperatures.

Into the second year and beyond, new casualties from radiation exposure would begin to appear, as later illnesses such as leukemia began. Simultaneously, increased internal doses of radiation could occur as people depleted pre-war food stores and consumed food contaminated by local and global fallout, often after bioconcentration. By then the climate would have returned to much less extreme conditions, allowing a renewal of primary productivity in terrestrial and aquatic ecosystems. Although agricultural productivity could recur (i.e., the physical environment would allow biotic production), the human support systems necessary for any but subsistence agriculture would not likely be operational. For most survivors, it could be expected that food shortages to the point of starvation would continue for many years into the future. It is entirely possible, perhaps likely, that globally the greatest mechanism for death associated with a large-scale nuclear war would be starvation.

Societal conditions and psychological states of survivors into the post-war period are extremely uncertain to predict. Controlling factors are likely to include the intense competition for resources, including food, water, shelter, and fuel. Reestablishment of organized societal systems on any but the local group scale could require very long periods of time, periods of continual conflict and lack of the cooperation necessary for redevelopment of economic, agricultural, transportation, communication, medical, and education systems. The lack of control by humans over their affairs would enhance exploitation of natural systems, as the carrying capacities based on human social systems far exceed the carrying capacities of unmanaged natural ecosystems. Because of this phenomenon, human recovery would likely proceed no more rapidly than the recovery of natural systems. Further, the human reliance on natural systems for support functions would likely retard the recovery of those systems.

The conditions described here are germane to the Northern Hemisphere, with respect to fallout radiation, light, temperature, and detonation effects. Depending on the rates of injection of particulates into stratospheric and/or trans-hemispheric atmospheric circulation patterns, the Southern Hemisphere could also experience substantial climatic alterations. Regardless, the human population of the Southern Hemisphere and other nontargeted countries would suffer widespread starvation as food imports were eliminated. No place on Earth could be relied upon to have benign environmental and societal conditions after a large-scale nuclear war.

The picture that emerges is stark and devastating. No refugia would exist. Total fatalities worldwide could be on the order of a few billion people. Because of the uncertainties in such estimates, the number cannot be resolved sufficiently finely to determine the likelihood of total deaths exceeding 5×10^9, i.e., for the total extinction of the human population. Nevertheless, for the first time in the millions of years of human biological and cultural evolution, there exists a single mechanism by which *Homo sapiens* could effect its own demise.

In summary, the consequence analyses here must include uncertainty. But that uncertainty in the effects on humans and biological systems is less than might be anticipated, since the stresses potentially imposed on living organisms are so extreme, so unprecedented in spatial extent, so abrupt in onset, so long-lasting in scope, and so fraught with synergisms and feedbacks, that the qualitative nature of a post-nuclear war can be reasonably projected. These projections suggest the following conclusions:

1. There is a real, nonzero possibility for there to be no human survivors in the Northern Hemisphere.
2. Depending on certain key atmospheric parameters, there is a potential to export that risk to the Southern Hemisphere. The

extinction of *Homo sapiens* is a valid scientific question, no longer merely the hyperbole of science fiction novelists.

3. If there were human survivors, they would be subject to extreme stress conditions for long periods of time. The optimal location to be if there is a nuclear war may well be at some ground zero.

4. The support functions of natural ecosystems would be severely curtailed, perhaps lost, at a time when human support functions would be terminated, and reliance would have to be placed on natural systems for human survival. Even absent stresses on those natural systems, they could not support the human population at current levels; how many people could be maintained by an extremely stressed Nature is highly uncertain.

5. Recovery, in the event that there were human survivors, would take a very long time and require the concomitant, linked recovery processes in human, ecological, and atmospheric systems. The legacy from a large-scale nuclear war would last beyond the foreseeable future, and nuclear war would constitute not just war on the combatant nations, but war on the environment itself and on all current and foreseeable future human generations.

References

Abrams, H. and W. Von Kaenal. 1981. Medical problems of survivors of nuclear war. *New Eng. J. Med*. 305: 1226–1232.

ACDA. 1979. *Effects of Nuclear War*. U.S. Arms Control and Disarmament Agency, Washington, D.C. 26 pp.

Adams, C.E., N.H. Farlow, and W.R. Schell. 1960. The compositions, structures and origins of radioactive fallout particles. *Geochimica et Cosmochimica Acta* 18:42–56.

Aleksandrov, V.V. and G.L. Stenchikov. 1983. On the modelling of the climatic consequences of nuclear war. USSR Academy of Sciences Computing Centre, Moscow.

Altman, P.L. and D.S. Dittmer (eds.). 1966. *Environmental Biology*. Federation of American Societies for Experimental Biology, Bethesda, MD. 694 pp.

Ambio Advisory Group. 1982. Reference scenario: how a nuclear war might be fought. *Ambio* 11: 94–99.

Anderson, H.E. 1970. Forest fuel ignitability. *Fire Technology* 6:312–319.

Archer, V. 1980. Effects of low-level radiation: a critical review. *Nuclear Safety* 21: 68–82.

Arkin, W., F. Von Hippel, and B.G. Levi. 1982. The consequences of a "limited" nuclear war in East and West Germany. *Ambio* 11: 163–173.

Armentano, T.V. and C.W. Ralston. 1980. The role of temperate zone forests in the global carbon cycle. *Can. J. For. Res*. 10:53–60.

Ashton, G.D. 1980. Freshwater ice growth, motion, and decay. *In*: Colbeck, S.C. *Dynamics of Snow and Ice Masses*. Academic Press, New York.

Ayres, R. 1965. *Environmental Effects of Nuclear Weapons*. Parts I, II, III. HI-518-RR. Hudson Research Institute, Harmon-on-Hudson, NY.

Bacastow, R.B. and C.D. Keeling. 1981. Atmospheric carbon dioxide concentration and the observed airborne fraction. *In*: B. Bolin (ed.) *Carbon Cycle Modelling*. SCOPE 16. John Wiley & Sons, New York.

Baker, D.N., J.D. Heskett, and W.G. Duncan. 1972. Simulation of growth and yield in cotton: I. Gross photosynthesis, respiration, and growth. *Crop Science* 12:431–435.

Barnaby, F. 1982. The effects of a global nuclear war: the arsenals. *Ambio* 11: 76–83.

Barnaby, F., and J. Rotblat. 1982. The effects of nuclear weapons. *Ambio* 11: 84–93.

Barr, T.N. 1981. The world food situation and global grain prospects. *Science* 214:1087–1095.

Beebe, G. 1982. Ionizing radiation and health—late-appearing effects of exposure. *Amer. Sci.* 70: 35–44.

Bell, G.H., D. Emslie-Smith, and C.R. Patterson. 1980. *Textbook of Physiology*. Longman, London.

Bensen, D. and A. Sparrow (eds.). 1971. *Survival of Food Crops and Livestock in the Event of Nuclear War*. U.S. Atomic Energy Commission, Washington, D.C. 754 pp.

Bergstrom, S., et al. 1983. *Effects of a Nuclear War on Health and Health Services*. Report of the International Committee of Experts in Medical Sciences and Public Health, WHO Pub. A36.12, Geneva, Switzerland.

Berry, J.A. and W.J.S. Downton. 1982. Environmental regulation of photosynthesis. *In*: Govindjee (ed.) *Photosynthesis. Vol. II. Development, Carbon Metabolism and Plant Productivity*. Academic Press, New York, London.

Billheimer, J.W. and A.W. Simpson. 1978. *Effects of Attack on Food Production to the Relocated Population. Vols. I and II*. SYSTAN, Inc., prepared for the U.S. Defense Civil Preparedness Agency, Washington, D.C.

Björkman, O. 1981. Responses to different quantum flux densities. *In*: O.L. Lange, P.S. Nobel, C.B. Osmond, and H. Ziegler (eds). *Encyclopedia of Plant Physiology. Vol. 12A Physiological Plant Ecology I. Responses to the Physical Environment*. Springer-Verlag, Berlin, Heidelberg, New York.

Bolin, B. 1977. Changes of land biota and their importance for the carbon cycle. *Science* 196:613–615.

Bolt, B. 1976. *Nuclear Explosions and Earthquakes*. W.N. Freeman, San Francisco, CA. 309 pp.

Bond, H. 1946. Fire casualties of the German attacks. *In*: H. Bond (ed.). *Fire and the Air War*. National Fire Protection Association, Boston, MA.

Bondietti, E. 1982. Effects of agriculture. *Ambio* 11: 138–142.

Bowen, R. 1982. *Surface Water*. Wiley-Interscience, New York. 290 pp.

Broido, A. 1965. Effects of fire on major ecosystems. *In*: G.M. Woodwell (ed.). *Ecological Effects of Nuclear War*. Brookhaven National Laboratory. BNL-917 (C-43). NTIS, Springfield, VA.

Bronowski, J. 1973. *The Ascent of Man*. Little, Brown, and Co., Boston. 448 pp.

Brown, L.R. 1981. World population growth, soil erosion, and food security. *Science* 214:995–1002.

Brown, S.L. and U.F. Pilz. 1969. *U.S. Agriculture: Potential Vulnerabilities*. Stanford Research Institute, prepared for the U.S. Defense Civil Preparedness Agency, Washington, D.C.

Brown, S.L., H. Lee, J.L. Mackin, and K.D. Moll. 1973. *Agricultural Vulnerability to Nuclear War*. Stanford Research Institute, prepared for the U.S. Defense Civil Preparedness Agency, Washington, D.C.

Burton, A.C. and O.G. Edholm. 1955. *Man in a Cold Environment*. Edward Arnold, Ltd., London. 273 pp.

Caldwell, M.M. 1981. Plant response to solar ultraviolet radiation. *In* O.L. Lange, P.S. Nobel, C.B. Osmond and H. Ziegler (eds.) *Encyclopedia of Plant Physiology. Vol. 12A Physiological Plant Ecology I. Responses to the Physical Environment*. Springer-Verlag, Berlin, Heidelberg, New York.

Caldwell, M.M., R. Robberecht, and W.D. Billings. 1980. A steep latitudinal gradient of solar ultraviolet-B radiation in the arctic-alpine life zone. *Ecology* 61:600–611.

Calkins, J. 1982. *The Role of Solar Ultraviolet Radiation in Marine Ecosystems*. Plenum Press, New York. 724 pp.

Center for Defense Information. 1982. *The Defense Monitor*. Washington, D.C.

Chandler, C.C., et al. 1963. Prediction of fire spread following nuclear explosion. U.S. Forest Service Res. Pap. PSW-5. Pacific Southwest Forest Range Experiment Station, Berkeley, CA.

Chazov, E. and M. Vartanian. 1982. Effects on human behavior. *Ambio* 11: 158–160.

Christy, A.L. and C.A. Porter. 1982. Canopy photosynthesis and yield in soybeans. *In*: Govindjee (ed.) *Photosynthesis. Vol. II. Development, Carbon Metabolism and Plant Productivity*. Academic Press, New York, London.

Coggle, J. and P. Lindop. 1982. Medical consequences of radiation following a global nuclear war. *Ambio* 11: 106–113.

Cohen, S. 1978. *The Neutron Bomb: Political, Technological, and Military Issues*. Institute for Foreign Policy Analysis, Cambridge, MA and Washington, D.C. 95 pp.

Colbeck, S.C. 1980. *Dynamics of Snow and Ice Masses*. Academic Press, New York. 468 pp.

Conrad, R., et al. 1975. *A Twenty-Year Review of Medical Findings in a Marshallese Population Accidentally Exposed to Radioactive Fallout*. U.S. Energy Research and Development Administration, Washington, D.C. 154 pp.

Covey, C., S.H. Schneider, and S.L. Thompson. 1984. Global atmospheric effects of massive smoke injections from a nuclear war: results from general circulation model simulations. *Nature* 308:21–25.

Cropper, W.P. and K.C. Ewel. 1983. Computer simulation of long-term carbon storage patterns in Florida slash pine plantations. *For. Ecol. Manage.* 6:101–114.

Crutzen, P.J. 1983. Atmospheric interactions—homogeneous gas reactions of C, N, and S containing compounds. *In*: B. Bolin and R.B. Cook (eds.) *The Major Biogeochemical Cycles and Their Interactions*. SCOPE. John Wiley & Sons, New York.

Crutzen, P. and J. Birks. 1982. The atmosphere after a nuclear war: twilight at noon. *Ambio* 11: 114–125.

Crutzen, P.J. and I.E. Galbally. 1984. Atmospheric effects from post-nuclear fires. *J. Atmospheric Chemistry* (in press).

Dahlman, O. and H. Israelson. 1977. *Monitoring Underground Nuclear Explosions*. Elsevier, New York. 440 pp.

DCPA. 1973. *DCPA Attack Manual*. Defense Civil Preparedness Agency, Department of Defense, CPG 2 - 1A1, Washington, D.C.

Edholm, O.G. 1974. Physiological research at British Antarctic Survey Stations. *In*: Gunderson, E. (ed.) *Human Adaptability to Antarctic Conditions, Vol. 22*. American Geophysical Union, Washington, D.C.

Edholm, O.G. 1978. *Man—Hot and Cold*. Edward Arnold, Ltd., London. 59 pp.

Edmonston, B. 1975. *Population Distribution in American Cities*. Lexington Books, Lexington, Massachusetts. 156 pp.

Edmonston, B. and T.M. Guterbock. 1984. Is suburbanization slowing down? Recent trends in population deconcentration in U.S. metropolitan areas. *Social Forces* 62(4):905–925.

Ehrlich, P.R., J. Harte, M.A. Harwell, P.H. Raven, C. Sagan, G.M. Woodwell, J. Berry, E.S. Ayensu, A.H. Ehrlich, T. Eisner, S.J. Gould, H.D. Grover,

R. Herrera, R.M. May, E. Mayr, C.P. McKay, H.A. Mooney, N. Myers, D. Pimentel, and J.M. Teal. 1983. Long-term biological consequences of nuclear war. *Science* 222:1293–1300.

Emmanuel, W.R., J.S. Olson, and G.G. Killough. 1980. The expanded use of fossil fuels by the U.S. and the global carbon dioxide problem. *J. Env. Mgt.* 10:37–49.

Erikson, K.T. 1976. Loss of communality at Buffalo Creek. *American Journal of Psychiatry* 133(3):302, cited in Chazev and Vartanian, 1982.

FAO. 1983. *The State of Food and Agriculture*. 1982. United Nations Food and Agriculture Organization, Rome.

FEMA. 1982a. *Food Vulnerability Briefing*. Unpublished briefing of the Cabinet Council by the Federal Emergency Management Agency, Washington, D.C.

FEMA. 1982b. *Attack Environment Manual*. CPG-2-1A2. Federal Emergency Management Agency, Washington, D.C.

Fetter, S. and K. Tsipis. 1981. Catastrophic releases of radioactivity. *Sci. Amer.* 244: 41–47.

Geiger, H.J. 1981. The illusion of survival. *Bulletin of Atomic Scientists* 37:16–20.

Geiger, H.J. 1983. Medical and public health problems of crisis relocation. *In*: J. Leaning and L. Keyes (eds.) *The Counterfeit Ark*. Ballinger, Cambridge, MA. 337 pp.

Glasstone, S. and P. Dolan. 1977. *The Effects of Nuclear Weapons* (3rd ed.) U.S. Government Printing Office, Washington, D.C. 653 pp.

Goffman, J.W. 1983. *Radiation and Human Health*. Sierra Club Books, San Francisco.

Grover, H.D., et al. (in prep.). Review of studies on the ecological consequences of nuclear war. Report to the Ecological Society of America.

Haaland, C.M., C.V. Chester, and E.P. Wigner. 1976. *Survival of the Relocated Population of the U.S. After a Nuclear Attack*. ORNL-5041, U.S. Defense Civil Preparedness Agency, Washington, D.C.

Hanson, J.D., J.W. Parton, and J.W. Skiles. 1983. SPUR: A rangeland model for management and research—the plant growth component. *In*: J.R. Wight (ed.). *SPUR—Simulation of Production and Utilization of Rangelands*. USDA ARS 1431, Washington, D.C.

Harwell, M.A. 1984. Feedbacks between human and ecological responses to nuclear war, pp. 118–124. *In*: P. Ehrlich, C. Sagan, D. Kennedy, and W.O. Roberts. *The Cold and the Dark*. *Proceedings of the Conference on the Long-Term Biological Consequences of Nuclear War*. W.W. Norton & Co., New York.

Harwell, M.A. and R.B. Stewart. 1984. The effects of nuclear winter on Canadian wheat and barley production. (In preparation.)

Harwell, M.A. and D.A. Weinstein. 1982. *Modelling the Effects of Air Pollution on Forested Ecosystems*. ERC Report 6. Ecosystems Research Center, Cornell University, Ithaca, NY.

Health and Welfare Canada. 1982. *Nitrogen Dioxide*. Minister of National Health and Welfare, Ottawa. 31 pp.

Hensel, H., K. Bruck, and P. Raths. 1973. Homeothermic organisms. *In*: H. Precht, J. Christophersen, H. Hensel, and W. Larcer (eds.) *Temperature and Life*. Springer-Verlag, Berlin, Heidelberg, New York.

Hersey, J. 1959. *Hiroshima*. Bantam Books, New York.

Hill, G. and P. Gardiner. 1979. *Managing the U.S. Economy in a Post-Attack Environment*. Federal Emergency Management Agency, Washington, D.C. 59 pp.

Hjort, H. 1983. The good Earth: agricultural productivity after nuclear war. *In*: J.

Leaning and L. Keyes (eds.) *The Counterfeit Ark*. Ballinger, Cambridge, MA. 337 pp.

Howell, F.G., J.B. Gentry, and M.H. Smith. 1975. *Mineral Cycling in Southeastern Ecosystems*. U.S. ERDA CONF-740513. NTIS, Springfield, VA.

Hutchinson, T.C. 1967. Comparative studies of the ability of species to withstand prolonged periods of darkness. *J. Ecol.* 55:291–299.

Ishikawa, E. and D. Swain. 1981. *Hiroshima and Nagasaki: The Physical, Medical, and Social Effects of the Atomic Bombings*. Basic Books, New York. 706 pp.

Johnson, D.G. 1983. *The World Grain Economy and Climate Change to The Year 2000: Implications for Policy*. National Defense University Press, Washington, D.C. 50 pp.

Katz, A.M. 1979. *Economic and Social Consequences of Nuclear Attacks on the United States*. Report to Committee on Banking, Housing, and Urban Affairs, U.S. Senate, Washington, D.C.

Katz, A. 1982. *Life After Nuclear War—The Economic and Social Impacts of Nuclear Attacks on the United States*. Ballinger, Cambridge, MA. 422 pp.

Kerr, J.W., et al. 1971. Nuclear weapons effects in a forest environment—thermal and fire (u). TTCP Panel N-2, Report No. 2:TR2-70. GE Tempo (DASIAC), Santa Barbara, CA.

Killian, H. 1981. *Disaster Medicine. Vol. 3. Cold and Frost Injuries*. Springer-Verlag, Berlin, Heidelberg, New York.

Knoll, E.,' and T. Postol. 1978. The day the bomb went off. *The Progressive* (October).

Kocher, D. 1981. *Radioactive Decay Tables—A Handbook of Decay Data for Application to Radiation Dosimetry and Radiological Assessments*. U.S. Department of Energy. DOE-TIC-11026. NTIS, Springfield, VA. 228 pp.

Kozlowski, T.T. and C.E. Ahlgren (eds.). 1974. *Fire and Ecosystems*. Academic Press, New York. 542 pp.

Kremer, J.N. and S.W. Nixon. 1978. *A Coastal Marine Ecosystem: Simulation and Analysis. Ecological Studies 24*. Springer-Verlag, Berlin, Heidelberg, New York. 215 pp.

Kress, I.W. and J.M. Skelly. 1982. Response of several Eastern forest tree species to chronic doses of ozone and nitrogen dioxide. *Plant Disease* 66:1149–1152.

Kuchler, A.S. 1964. *Potential Natural Vegetation of the Conterminous United States*. Am. Geographical Soc. Spec. Pub. No. 36, Washington, D.C.

Larcher, W. 1980. *Physiological Plant Ecology* (2nd ed.) Springer-Verlag, Berlin, Heidelberg, New York.

Larcher, W. and H. Bauer. 1981. Ecological significance of resistance to low temperature. *In*: O.L. Lange, P.S. Nobel, C.B. Osmond, and H. Ziegler (eds.) *Encyclopedia of Plant Physiology. Vol. 12A Physiological Plant Ecology I. Responses to the Physical Environment*. Springer-Verlag, Berlin, Heidelberg, New York.

Laulan, Y. 1982. Economic consequences: back to the dark ages. *Ambio* 11: 149–152.

Leaning, J. 1982. *Civil Defense in the Nuclear Age*. Physicians for Social Responsibility, Cambridge, MA.

Leaning, J. 1983. An ill wind: Radiation consequences of nuclear war. *In*: J. Leaning and L. Keyes (eds.). *The Counterfeit Ark*. Ballinger, Cambridge, MA.

Leaning, J. and L. Keyes (eds.). 1983. *The Counterfeit Ark*. Ballinger, Cambridge, MA. 337 pp.

LeBlanc, J. 1975. *Man in the Cold*. C.C. Thomas, Publishers, Springfield, IL. 195 pp.

LeBlanc, J. 1978. Adaptation of man to cold. In: Wang, L. and J. Hudson (eds.). *Strategies in Cold: Natural Torpidity and Thermogenesis*. Academic Press, New York.

Levin, S.A., et al. 1983. *New Perspectives in Ecotoxicology*. ERC Report 14. Ecosystems Research Center, Cornell University, Ithaca, NY.

Levitt, J. 1980. *Responses of Plant to Environmental Stresses*. Academic Press, London, New York.

Lewis, K. 1979. The prompt and delayed effects of nuclear war. *Sci. Amer.* 241: 35–47.

Lifton, R.J. 1967. *Death in Life: Survivors of Hiroshima*. Random House, New York.

Lifton, R.J., E. Markusen, and D. Austin. 1983. The second death: psychological survival after nuclear war. *In*: J. Leaning and L. Keyes (eds.) *The Counterfeit Ark*. Ballinger, Cambridge, MA. 357 pp.

Lundquist, S. 1983. Electromagnetic pulse effects from a nuclear war. Unpublished presentation to SCOPE conference on the environmental effects of nuclear war, Stockholm, Sweden.

Lynch, K. 1983. Coming home: the urban environment after nuclear war. *In*: J. Leaning and L. Keyes (eds.) *The Counterfeit Ark*. Ballinger, Cambridge, MA. 337 pp.

MacCracken, M.C. 1983. *Nuclear War: Preliminary Estimates of the Climatic Effects of a Nuclear Exchange*. UCRL-89770, Lawrence Livermore National Laboratory, Livermore, CA.

Malone, T. 1974. Transcript of National Academy of Sciences luncheon meeting, 28 May. Washington, D.C.

Mark, J.C. 1976. Global consequences of nuclear weaponry. *Ann. Rev. Nuclear Science* 26: 51–87.

Martin, S.B. 1974. *The Role of Fire in Nuclear Warfare*. URS Research Co., San Mateo, CA. Defense Nuclear Agency 2692F.

Middleton, H. 1982. Epidemiology: the future is sickness and death. *Ambio* 11:100–105.

Miller, D.H. 1977. *Water at the Surface of the Earth*. Academic Press, New York. 557 pp.

Miller, J. 1982. Postal service plan covers snow, sleet and atom war. *New York Times*, 13 Aug. 1982, p. A26.

Mooney, H.A., J.R. Ehleringer, and J.A. Berry. 1976. High photosynthetic capacity of a winter annual in Death Valley. *Science* 194:322–333.

Mooney, H.A., C. Field, and C. Vazques-Yanes. 1984. Photosynthetic characteristics of wet tropical forest plants. *In: Physiological Ecology of Plants of the Wet Tropics:* 113–128. Junk, The Hague.

Moore, B. et al. 1981. A simple model for analysis of the role of terrestrial ecosystems in the global carbon budget. *In*: B. Bolin (ed.). *Carbon Cycle Modelling*. SCOPE 16. John Wiley & Sons, New York.

Morland, H. 1979. The H-bomb secret. *The Progressive* (November): 14-23.

Mullaney, A.J. 1946. German fire departments under air attack. *In*: H. Bond (ed.). *Fire and the Air War*. National Fire Protection Association, Boston, MA.

Murray, R.C. and E.B. Reeves. 1977. *Estimated Use of Water in the United States in 1975*. U.S. Geol. Surv. Circ. 765, Washington, D.C.

Naidu, J. 1984. Potential effects on groundwater following a nuclear attack. Unpublished presentation at SCOPE conference on the consequences of nuclear war, New Delhi, India.

NAS. 1972. *The Effects on Populations of Exposure to Low Levels of Ionizing Radiation*.

(BEIR I) National Academy of Sciences, Washington, D.C. 217 pp.

NAS. 1975. *Long-term Worldwide Effects of Multiple Nuclear-Weapons Detonations*. National Academy of Sciences, Washington, D.C. 213 pp.

NAS. 1977. *Committee on Medical and Biological Effects of Environmental Pollutants Nitrogen Oxides*. National Academy of Sciences, Washington, D.C.

NAS. 1980. *The Effects on Populations of Exposure to Low Levels of Ionizing Radiation*. (BEIR III) National Academy of Sciences, Washington, D.C. 628 pp.

NAS. 1981. *Effect of Environment on Nutrient Requirements of Domestic Animals*. National Academy of Sciences, Washington, D.C. 152 pp.

NAS. 1982. *Causes and Effects of Stratospheris Ozone Reduction: An Update*. National Academy of Sciences, Washington, D.C. 399 pp.

NAS. (in prep.). *Report of the Committee on the Atmospheric Effects of Nuclear Explosions*. National Academy of Sciences, Washington, D.C.

Neumann, J. 1959. Maximum depth of lakes. *Journal of the Fisheries Research Board of Canada*. 16:923–927.

OTA. 1979. *The Effects of Nuclear War*. U.S. Congress, Office of Technology Assessment, Washington, D.C. 151 pp.

Penman, H.L. 1970. The water cycle. *Sci. Amer.* 223(3):99–108.

Pimentel, D. (ed.) 1980. *Handbook of Energy Utilization in Agriculture*. CRC Press, Boca Raton, FL. 475 pp.

Pimentel, D. and M. Burgess. 1980. Energy inputs in corn production. *In*: D. Pimentel (ed.) *Handbook of Energy Utilization in Agriculture*. CRC Press, Boca Raton, FL. 475 pp.

Pimentel, D. and M. Pimentel. 1979. *Food, Energy and Society*. Edward Arnold Ltd., London. 165 pp.

Pimentel, D., P.A. Oltenacu, M.C. Nesheim, J. Krummel, M.S. Allen, and S. Chick. 1980a. Grass-fed livestock potential: energy and land constraints. *Science* 207:843–848.

Pimentel, D., E. Garnick, A. Berkowitz, S. Jacobson, P. Black, S. Valdes-Cogliano, B. Vinzant, E. Hudes, and S. Littman. 1980b. *Environmental Quality and Natural Biota*. Environ. Biol. Rept. 80-1, Cornell University, Ithaca, NY.

Pitts, D.M. 1983. Testimony, pp. 83-101. *In: Hearing on the Consequences of Nuclear War on the Global Environment*. U.S. Congress, Washington, D.C.

Precht, H. 1973. Limiting temperature of life functions. In: H. Precht, J. Christopherson, H. Hensel, and W. Larcher (eds.) *Temperature and Life*. Springer-Verlag, Berlin, Heidelberg, New York.

Pritchard, D.W., et al. 1971. *Radioactivity in the Marine Environment*. National Academy of Sciences, Washington, D.C.

Ramberg, B. 1980. *Destruction of Nuclear Energy Facilities in War—The Problem and the Implications*. Lexington Books, Lexington, MA. 203 pp.

Rand McNally. 1983. *Commercial Atlas and Marketing Guide*. 114th ed. Rand McNally, New York.

Ream, C.H. 1981. The effects of fire and other disturbances on small mammals and their predators: an annotated bibliography. USDA Forest Service Gen. Tech. Rep. INT-106. Intermountain Forest and Range Experiment Station, Ogden, UT.

Robock, A. 1984. Nuclear winter: Energy balance climate model finds large longer-lasting effects. *Nature* (in press).

Rotblat, J. 1981. *Nuclear Radiation in Warfare*. Stockholm International Peace Research Institute. Oelgeschlager, Gunn, and Hain, Cambridge, MA. 149 pp.

Rotty, R.M. 1981. Data for global CO_2 production from fossil fuels and cement.

In: B. Bolin (ed.). *Carbon Cycle Modelling*. SCOPE 16. John Wiley & Sons, New York.

Sagan, C. 1983. Nuclear war and climatic catastrophe: some policy implications. *Foreign Affairs* 62:257–292.

Schell, J. 1982. *The Fate of the Earth*. Knopf, New York. 244 pp.

Setlow, R.B. 1974. The wavelength of sunlight effective in producing skin cancer: a theoretical analysis. *Proc. Natl. Acad. Sci. (USA)* 71:3363–3366.

Seymour, A. 1982. The impact on ocean ecosystems. *Ambio* 11: 132–137.

Siemes, Fr. 1946. Hiroshima—August 6, 1945. *Bulletin of the Atomic Scientists*, cited in Katz, 1982.

SIPRI. 1977. *Weapons of Mass Destruction and the Environment*. Crane, Russak, New York. 95 pp.

SIPRI. 1980. *Warfare in a Fragile World—Military Impact on the Human Environment*. Crane, Russak, New York. 249 pp.

SIPRI. 1982a. *Agreements for Arms Control: A Critical Survey*. Taylor and Francis, London. 387 pp.

SIPRI. 1982b. *The Arms Race and Arms Control*. Taylor and Francis, London. 242 pp.

SIPRI. 1982c. *Outer Space—A New Dimension of the Arms Race*. Taylor and Francis, London. 423 pp.

SIPRI. 1982d. *World Armaments and Disarmament—SIPRI Yearbook 1982*. Taylor and Francis, London. 517 pp.

Smith, W.H. 1981. *Air Pollution and Forests*. Springer-Verlag, Berlin, Heidelberg, New York. 379 pp.

Sokolov, A.A. and T.G. Chapman (eds.). 1974. *Methods for Water Balance Computations*. UNESCO Press, Paris. 127 pp.

Strope, W.E. and J.F. Christian. 1964. *Fire Aspects of Civil Defense*. OCD-TR-25. Office of Civil Defense, Washington, D.C.

Tamrazyan, G.P. 1974. Total lake water resources of the planet. *Bulletin of the Geological Society of Finland*. 46:23–27.

Teramura, A.H., R.H. Biggs, and S. Kossuth. 1980. Effects of ultraviolet-B irradiances on soybean. II. Interaction between ultraviolet-B and photosynthetically active radiation on net photosynthesis, dark respiration, and transpiration. *Plant Physiology* 65:483–488.

Treshow, M. and D. Stewart. 1973. Ozone sensitivity of plants in natural communities. *Biological Conservation* 5:209–214.

Turco, R.P., O.B. Toon, T.P. Ackerman, J.P. Pollack, and C. Sagan. 1983a. Nuclear winter: global consequences of multiple nuclear explosions. *Science* 222:1283–1292.

Turco, R.P., O.B. Toon, T.P. Ackerman, J.P. Pollack, and C. Sagan. 1983b. Long-term atmospheric and climatic consequences of a nuclear exchange. Unpublished manuscript.

Turco, R.P., O.B. Toon, T.P. Ackerman, J.P. Pollack, and C. Sagan. 1984. The climatic effects of nuclear war. *Sci. Amer.* 251:33–43.

UN. 1968. *Effects of the Possible Use of Nuclear Weapons and the Security and Economic Implications for States of the Acquisition and Further Development of These Weapons*. United Nations, New York. 76 pp.

UN. 1979. *The Effects of Weapons on Ecosystems*. United Nations Environmental Programme, New York. 70 pp.

UN. 1980. *Nuclear Weapons*. Autumn Press, Brookline, MA. 223 pp.

UN. 1981. *Comprehensive Study on Nuclear Weapons*. United Nations, Report of the Secretary General, New York. 172 pp.

U.S. Bureau of the Census. 1974. *Urban Atlas*. Tract Data for Standard Metropolitan Areas, Anaheim-Santa Ana-Garden Grove, California; Boston, Massachusetts; Chicago, Illinois; Dallas, Texas; Detroit, Michigan, Fort Worth, Texas; Houston, Texas; Los Angeles-Long Beach, California; New York, New York; Philadelphia, Pennsylvania-New Jersey; San Bernadino-Riverside-Ontario, California; San Francisco-Oakland, California; Washington, D.C. U.S. Department of Commerce, Washington, D.C.

U.S. Bureau of the Census. 1980. 1980 *Maps of County Subdivisions and Places*: California (North and South), Illinois, Indiana, Maryland-Delaware, Massachusetts-Connecticut-Rhode Island, Michigan, New Jersey, New York, Pennsylvania, Texas, Virginia. U.S. Department of Commerce, Washington, D.C.

U.S. Bureau of the Census. 1981. *Statistical Abstract of the United States 1981*. 102nd ed. U.S. Department of Commerce, Washington, D.C.

U.S. Bureau of the Census. 1982. *1980 Census of Population*: Number of Inhabitants, Part 6, California; Part 8, Connecticut; Part 9, Delaware; Part 15, Illinois; Part 16, Indiana; Part 22, Maryland; Part 23, Massachusetts; Part 24, Michigan; Part 32, New Jersey; Part 34, New York; Part 40, Pennsylvania; Part 45, Texas; Part 48, Virginia. U.S. Department of Commerce, Washington, D.C.

U.S. Bureau of the Census. 1983. *1980 Census of Population*: Number of Inhabitants, Part 1, U.S. Summary. U.S. Department of Commerce, Washington, D.C.

U.S. Congress. 1982. *The Consequences of Nuclear War on the Global Environment*. Hearing before Committee on Science and Technology, U.S. House of Representatives, Washington, D.C.

U.S. Department of Interior. 1980. *The World's Major Dams, Man-Made Lakes, and Hydroelectric Plants*. Office of Science and Technology, Washington, D.C.

U.S. Strategic Bombing Survey. 1946. The effects of atomic bombs on Hiroshima and Nagasaki. *In*: H. Bond (ed.). *Fire and the Air War*. National Fire Protection Association, Boston, MA.

USDA. 1980. *Agricultural Statistics 1980*. U.S. Government Printing office, Washington, D.C.

USDA. 1981. *Agricultural Statistics 1981*. U.S. Government Printing Office, Washington, D.C.

USDA. 1982. *World Food Aid Needs and Availabilities, 1982*. U.S. Department of Agriculture, Washington, D.C.

USSR Committee for the International Hydrologic Decade. 1977. *Atlas of World Water Balance*. UNESCO Press, Paris.

USSR Committee for the International Hydrologic Decade. 1978. *World Water Balance and Water Resources of the Earth*. UNESCO Press, Paris. 663 pp.

Warner, C.W. and M.M. Caldwell. 1983. Influence of photon flux density in the 400–700 nm waveband on inhibition of photosynthesis by UV-B (280-320 nm) irradiation in soybean leaves: Separation of indirect and immediate effects. *Photochemistry and Photobiology*.

Weinstein, D.W. 1983. *A User's Guide to FORNUT*. ERC Report No. 28, Ecosystems Research Center, Cornell University, Ithaca, NY.

Weinstein, D.W. and M.A. Harwell. 1984. Modelling the potential for forest growth alteration as a consequence of gaseous pollutants and acidic precipitation. *Bulletin of Ecological Society of America* (in press).

Weinstein, D.W., M.A. Harwell, and B.L. Bedford. 1983. Detecting forest productivity reductions due to atmospheric pollutants. *In*: J.S. Jacobson and

L.S. Raymond, Jr. (eds.). *Proceedings 2nd New York State Symposium on Atmospheric Deposition*. Center for Environmental Research, Cornell University, Ithaca, New York.

Westing, A. 1978. Military impact on ocean ecology. *In*: Borgese, M. and N. Ginsburg (eds.). *Ocean Yearbook 1*. Univ. of Chicago Press, Chicago.

Wetzel, K.G. 1982. Effects on global supplies of freshwater. *Ambio* 11: 126–131.

Wetzel, R.G. 1975. *Limnology*. W.B. Saunders Co., Philadelphia. 743 pp.

Whittaker, R.H. and G.E. Likens. 1973. Carbon in the biota. *In*: G.M. Woodwell and E.V. Pecan (eds.). *Carbon and the Biosphere*. U.S. AEC CONF-720510, NTIS, Springfield, VA.

Wiersma, S.J. and S.B. Martin. 1973. *Evaluation of the Nuclear Fire Threat to Urban Areas*. Defence Civil Preparedness Agency, Washington, D.C.

Wight, J.R. (ed.). 1983. *SPUR—Simulation of Production and Utilization of Rangelands: A Rangeland Model for Management and Research*. USDA ARS 1431, Washington, D.C. 120 pp.

Wight, J.R., E.P. Springer and D.L. Brakensick. 1983. SPUR: a rangeland model for management and research—model overview. *In*: J.R. Wight (ed.). *SPUR— Simulation of Production and Utilization of Rangelands*. USDA ARS 1431, Washington, D.C.

Williams, G.P. 1965. Correlating freeze-up and break-up with weather conditions. *Canadian Geotechnical Journal* 2(4):313–326.

Woodmansee, R.G. and L.S. Wallach. 1981. Effects of fire regimes on biogeochemical cycles. *In*: Mooney, H., et al. (eds.). *Fire Regimes and Ecosystem Properties*. USDA Forest Service Gen. Tech. Rep. WO-26, Washington, D.C.

Woodwell, G.W. (ed.). 1963. *The Ecological Effect of Nuclear War*. Brookhaven National Laboratory. BNL-917 (C-43). NTIS, Springfield, VA.

Woodwell, G. 1982. The biotic effects of ionizing radiation. *Ambio* 11: 143–148.

Wright, H.A. and A.W. Bailey. 1982. *Fire Ecology*. John Wiley and Sons, New York. 501 pp.

Zelitch, I. 1982. The close relationship between net photosynthesis and crop yield. *BioScience* 32:796–802.

Zuckerman, S. 1982. *Nuclear Illusion and Reality*. Viking Press, New York.

Index